Crop Improvement in Aromatic Plants

NIPA® GENX ELECTRONIC RESOURCES & SOLUTIONS P. LTD.
New Delhi-110 034

About the Authors

Dr. Pritam Kumar Verma, Former Professor, Department of Genetics & Plant Breeding, Chaudhary Charan Singh Haryana Agricultural University, Hisar (Haryana) obtained his B.Sc. (Hons.) Agriculture and M.Sc. (Agri.) in the discipline of Plant Breeding from the then Haryana Agricultural University Hisar during 1979 and 1981, respectively. He obtained his Ph. D. (Agri.) in Plant Breeding degree from the then Himachal Pradesh Krishi Vishwavidyalaya, Palampur (H.P.) during 1993 as an in-service candidate having Senior Research Fellowship of ICAR. Professor Verma joined as Assistant Scientist in February, 1982; and subsequently promoted as Sr. Scientist and Principal Scientist/Prof. in Feb., 1995 and 2003, respectively. He has guided five M.Sc. and two Ph. D. students pursuing for degree in Genetics & Plant Breeding discipline. He has published about 100 research papers, including popular articles and abstracts on different crops, while working on Medicinal & Aromatic Plants, Rice, Wheat, Pearl millet, Mung bean, Castor Sesame, Chickpea, Raya etc. He has been associated in the ICAR Team Award for Wheat Improvement for the biennial 1994-96. He has organized and compiled four proceedings of National Seminars & Conferences on Medicinal & Aromatic Plants sponsored by National Medicinal Plants Board, New Delhi; Directorate of Arecanut & Spices Development, Calicut (Kerala) during 2002-2009; and Winter School on "Advances in Medicinal, Aromatic and Under-utilized Plants Research" sponsored by Indian Council of Agricultural Research, New Delhi during Sept.-Oct., 2005. He worked for 21 years on Medicinal & Aromatic Plants in different capacity. Besides, he worked for about eight years in Dryland Agriculture Unit in AICRPDA under CRIDA (ICAR), Hyderabad (Telangana) on various crops suited to dryland/rainfed agriculture; two years at Rice Research Station, Kaul (Kurukshetra); one year at Cotton Research Station, Sirsa; and finally as Head, Forage Section, Department of Genetics & Plant Breeding, CCS HAU, Hisar working on different crops during his service tenure of 37 years. He has been associated in the development of 11 crop varieties before retiring in November, 2018.

Dr. Rajesh Kumar Arya, presently working as Sr. Scientist & Head (MAP Section) obtained his M.Sc. (Agri.) in Plant Breeding during 2002 and Ph.D. (Agri.) in Plant Breeding during 2007 from the Department of Genetics and Plant Breeding CCS Haryana Agricultural University, Hisar. He has 20 years experience of research, including last 8 years on Medicinal, Aromatic and Potential crops for development of new varieties for higher yield with better quality of bioactive compounds. He has been awarded Appreciation Award in 2023 for his outstanding research by IFSDAA, Gottingen, Germany. He also received RMSI best paper award from Range Management and Agro-forestry Society of India in 2019. He has been presently associated in teaching PG course on Breeding for Medicinal and Aromatic plants. He has published more than 200 research papers in Journals of International and National repute. He has also published four books and three booklets and 10 book chapters. He is working as Editorial Secretary in journal on Forage Research; and Associate Editor in Ekin J. Crop Breeding and Genetics. He has developed two varieties of Fababean, one of Chandersur, six of Wheat and two of Ashwagandha for commercial cultivation at National level. He has been member of "Society for Sustainable Agriculture and Resource Management (SSARM)", India; and "International Foundation for Sustainable Development" in Africa and Asia (IFSDAA), Germany.

Crop Improvement in Aromatic Plants

P.K. Verma
Former Professor and Head
Medicinal, Aromatic & Potential Crops Section
Department of Genetics and Plant Breeding
CCS Haryana Agricultural University
Hisar, Haryana

R.K. Arya
Sr. Scientist & Head
Medicinal, Aromatic & Potential Crops Section
Department of Genetics and Plant Breeding
CCS Haryana Agricultural University
Hisar, Haryana

NIPA® GENX ELECTRONIC RESOURCES & SOLUTIONS P. LTD.
New Delhi-110 034

NIPA® GENX ELECTRONIC
RESOURCES & SOLUTIONS P. LTD.

101,103, Vikas Surya Plaza, CU Block
L.S.C. Market, Pitam Pura, New Delhi-110 034
Ph : +91-11-43860225, Mob.: +91 9717133558, 9540816132
E-mail: newindiapublishingagency@gmail.com
Website: www.nipaersources.com

Print ISBN: 978-93-58873-58-0
ebook ISBN: 978-93-58874-63-1

Composed and Designed by NIPA®.

Dedicated to
Our Children
Nisha & Kavita Verma
Arjoo & Alok Arya

Prof. B. R. Kamboj
Vice Chancellor
CCS HAU, Hisar (Haryana)

Foreword

It is a matter of immense pleasure and proud privilege for me to know that Dr. Pritam Kumar Verma, former Professor (Genetics & Plant Breeding) and Dr. Rajesh Kumar Arya, Sr. Scientist & Head, Medicinal, Aromatic and Potential Crops Section, Department of Genetics and Plant Breeding, CCS HAU, Hisar have prepared a text book on "Crop Improvement in Aromatic Plants". This will be the first book of its own kind on this aspect, which is need of the hours to stride ahead a step for the development of Aromatic Plants Sector in India.

India is one of the 12 mega biodiversity centres in the world, accounting for about 8% of plant biodiversity. Keeping this in view, India comprises of 16 different agro-climatic zones endowed with rich wealth of plant biodiversity and divided into eight regions having rich wealth of Medicinal and Aromatic plants. In-spite of this, majority of the produce pertaining to these plants comes from wild sources in forests, threatening the loss of their existence in near future due to over exploitation without conservation and replacement with the new plantation.

The cosmetic and perfumery industries are dependent on the production of aromatic plants, known for essential oils extracted from their various parts. Besides, their utility as fragrance, flavour and aroma for manufacturing of scents and other cosmetic items. These plants are of great utility as bio-pesticides, fungicides, nematicides and high valued processed products in the form of aromatherapy, analgesic creams, mouthwash liquids and immunity booster medicines.

Therefore, this text will be proven as mile stone in the development of sector dependent on the aromatic plants by way of providing basic knowledge to the scientists engaged in research and interested students perusing for higher education in the discipline of Genetics & Plant Breeding.

Hence, I convey my best wishes for the success and congratulate the scientists who have put up their best efforts in this endeavour while preparing this book.

(B. R. Kamboj)

Preface

This book has been prepared as a text-book, keeping in view of the post graduate students of State Agricultural Universities (SAUs) of India and abroad perusing for M.Sc. (Agri.) and Ph.D. (Agri.) in the discipline of Crop Improvement/ Genetics and Plant Breeding. An attempt has been made to present the basic requirements needed for perusing the crop improvement/breeding programme on the development of aromatic plants taking into consideration of origin nature of crop, availability of genotypic variation with regard to availability of different species within genus and between genus.

The text book has been written keeping in view of the courses offered during post graduate programme approved by the council (ICAR) for agricultural universities across the country. Important and necessary aromatic plant crops have been discussed and presented in this book with all the basic requirements viz. genotypic variability, genetic variability, cytogenetics, availability of presently released varieties from different institutions in India, breeding methodology to be used or useful, polyploidy, mutation, hybridization, clonal selection and biotechnological approaches.

Keeping in view, the scattered and non-availability of compiled information on the aspects of crop improvement/genetic & plant breeding in the literature, first attempt has been made in this regard for the benefits of the students perusing for their degree and the teachers/scientists engaged in teaching for the betterment/improvement of aromatic plants for their higher economic end produce and quality of essential oil for major constituents required for the industries engaged in cosmetic, perfumery, aromatherapy and pharmaceuticals developing analgesic, mouth-freshness and immunity booster medicines using different aromatic plants extracts.

In this attempt, help have been received from our colleagues of the university working in this direction in one or the other in the development of this text. Literature and references have been collected from the earlier work done at Chaudhary Charan Singh Haryana Agricultural University, Hisar and other SAUs, institutes of ICAR & CSIR and through internet.

Efforts have also been made to keep the text book as simple as possible to make it understood by everyone undergone through it. The usefulness and quality of this text book, but the errors, omissions and deficiencies are our own copyright and no one else should be credited for them.

In the end, we are grateful and express appreciation for Late Mrs. Verma and Late Mrs Arya; children and grand children, who have been affectionate, understanding and patient during the long time we spent while preparing this manuscript.

Hisar, 2025

P.K. Verma
Former Prof. (G&PB)
R.K. Arya
Sr. Scientist & Head (MAP Section)

Contents

1

Significance of Crop Improvement in Aromatic Plants

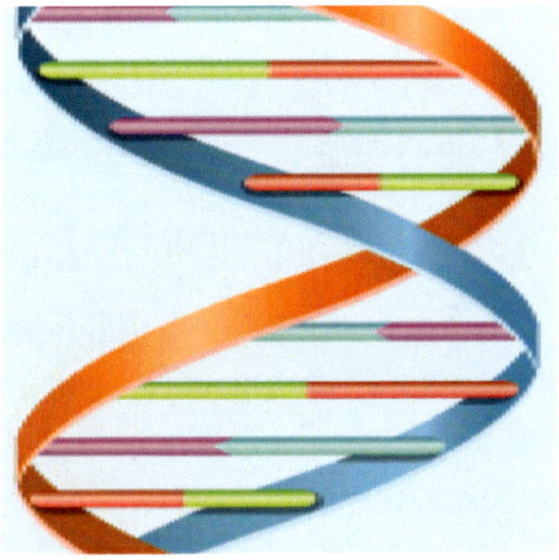

Aromatic plants are the life line of essential oil trades all over the world. India is considered to be botanical garden on account of prevalence of rich plant biodiversity owing to varied topography and wide range of climatic conditions present in different parts of the country. About 1300 plant species have been reported to be of aromatic value and 80 plant species are commercially grown and in high demand, but only 42 plant species are being exploited by modern pharmaceutical and essential oil industries as a source of raw materials for a range of natural perfumes, flavours, cosmetics, food and flavouring and aroma–therapy.

Plant-genetic diversity is the key component of agricultural production system, and no natural evolutionary adjustment of the system to changing environment and biotic conditions would be possible without it. India is richly endowed with plant wealth, utilized either directly for cultivation or as source of useful traits for breeding improved varieties. Collection, conservation and sustainable use of available gene pool and their wild relatives require specific attention keeping in view their overall importance and significance in agriculture and food security vis-à-vis day–to-day needs.

Cultivation of aromatic plants having aroma due to presence of an essential oil in their various parts needs to be made competitive against other seasonal crops, so that production from cultivated harvests are preferred over collections from the wild. There are possibilities of using the produce in cosmetic, perfumery and aromatherapy industries of several aromatic plants. So, the breeding strategy of aromatic crops should aim for improving the contents of therapeutic compounds, yield of grains, biomass yield, roots yield and oil quality & quantity. Rose essential oil is one of the most precious fragrance materials. Indian contribution to world supply of rose essential oil is minimal due to relative cost of production being high. Indian varieties of essential oil producing rose come to flower in a brief season at the end of winter in the Indo-Gangetic plains. Varieties of rose that may flower year round need to be developed by recombining suitable traits found in different kind of wild and cultivated roses.

Genetic improvement for developing superior cultivars through plant breeding approaches confers more than one advantages for aromatic plants ensures:

- Fixing high genetic ceiling for productivity.
- Introducing impressive uniformity of crop attracting grower's acceptance.
- Improving quality and homogeneity of produce.
- Increasing cost effectiveness by using improved varieties under cultivation.
- Value addition by developing specialty chemotypes.
- Facilitating diversification.
- Conserving genetic resources *in-vivo* and or *in-vitro.*

Thus, improved varieties are endowed with high yield potential for raw materials of high quality, impressive genetic uniformity of the crop, value addition, reduced cost of cultivation and wide adaptability for diverse biotic and abiotic stresses, and also for intensive cropping system. In view of these points, scientists have developed a large number of improved cultivars of important aromatic plants for their adaptive capacity, yield potential and quality components. Adoption of improved varieties of aromatic pants developed by State Agriculture Universities and ICAR institutes provide sustainable utilization of National Plants Genetic Wealth. Improved varieties of aromatic plants are adopted for:

- Cost-effective cultivation
- Total quality management (TQM)
- Conserving the National Plants Genetic Resources (PGR)

The plant breeder makes several decisions before starting any crop breeding programme in order to improve yield, adaptability and quality. Therefore, selection of parental plants from the available genetic resources, hybridization between them to select for desirable ones and what to discard & keep. The decision depends on the plants product of interest, available gene pool in related and wild species for improving yield, good quality, resistance to pests & diseases, tolerance to biotic & abiotic stresses and stability for wider adaptation.

The breeding objectives for aromatic plants are more directed towards the increase of biomass of the herb yield, from which essential oils are extracted. For example, breeding for vetiver/khas (*Vetiveria zizanioides*) is directed towards the increase of root biomass. The breeding objectives of aromatic grasses *viz.* Palmarosa, Lemongrass, Citronella grass (*Cymbopogon* spp.) are increased herbage and essential oil yields, higher concentration of desired chemical constituents in the oil, adaptation to wide agro-climatic conditions and resistance to biotic and abiotic stresses.

However, in the breeding of aromatic plants, therapeutic value is most important factor in addition to the higher yield. In aromatic plants, the inheritance of important chemical constituents is more important, as these are to be used as the product for aroma in cosmetic and perfumery industries. The morphological characters may be misleading for aromatic plants unless these characters are correlated with the yield and quality of aromatic compounds.

Conventional breeding approach

Genetic resources are building block of any breeding programme for improvement. Chomchalow (1980) reported the importance of genetic resources in the improvement of aromatic plants. Before starting any breeding programme, plant breeder selects and combines useful characters by hybridization to produce a desired variety. When plant breeders have exploited most of the available genetic variability in a species, they look for other sources of variability through hybridization with related or wild species.

Although the cultivation of aromatic plants has been known for centuries, but their germplasm collection, evaluation and utilization in breeding is very limited. It has already been emphasized in the FAO, 1993 report that germplasm collection and conservation is an urgent requirement for genetic erosion of aromatic plants. The deforestation and urbanization has resulted for the loss of many valuable wild species in the region. The traditional breeding methods for the improvement of aromatic plants involve evaluation, recombination through hybridization and selection.

The evaluation of contents of active principles in a given environment is a pre-requisite for aromatic plant breeding. The aromatic industries give priority to high quality from a small number of plants over the higher yield from a large number of plants. Therefore, breeder should stress increasing accumulation and yield of active principles rather than selecting for morphological differences.

Biotechnological approach

A plant breeder must work with the genetic variability available in parental material, which is more or less locally adapted. He attempts crosses in order to generate recombinants, which transgress with respect to the better parent. The amount of recombination in one breeding cycle depend on closeness of the parents, breeding system, propagation system and population size. Hybridization has been carried out in many aromatic plants to generate heterotic materials, so that selections can be made.

Wide/distant hybridization

Significant results can be obtained from**,** which involves locally adapted and non-adapted or wild parents. The success of the cross enables breeder to incorporate desired genes through introgression. In this method, later generations can differ in quality properties than parents. The other disadvantage is hybrid sterility, but this can be overcome by vegetative propagation or by producing polyploids. The use of **polyploids** in improvement of aromatic plant has been successful with many crops (Tetenyi 1991).

Chromosome manipulation

It is another technique used to improve the yield of aromatic plants. **Aneuploids** can be used to locate genes on the particular chromosomes to establish linkage group and for substitution of chromosome. If the result of fertilization of gametes containing non-separated homologus chromosome pairs and a normal gamete is viable, then **trisomic and monosomic plants** can be obtained.

Mutation

Induced mutation could be used in the improvement of plant species. Plants can be treated with radiation of a physical or chemical nature. The radiation treatment can provide a large variation for selection, which may encourage industries to use the materials. It can also cause **heterosis**. Mutants were also produced by radiation treatment of cell cultures, and variation in quality traits was recorded in regenerated plants.

It is evident from the literature that the traditional breeding strategies have been successful in obtaining varieties or cultivars with increased yield, wider agro-climatic adaptation, resistance to disease, pests and tolerance to stress. Though plant breeders find difficulties in breeding quantitative characters, yet some of the difficulties can be overcome by combining the traditional and fast-growing molecular techniques. This will enable both traditional and molecular breeders to work together to respond quickly and effectively to the constantly changing needs of crop improvement. These techniques can provide genes to traditional and molecular breeders to work together to respond quickly and effectively to the constantly changing needs of crop improvement. These techniques can provide genes to traditional plant breeders from previous inaccessible organisms (e.g. viruses, bacteria, algae, animals and unrelated plants) bypassing the normal sexual reproductive process, by transferring isolated genes and by rescuing to hybrid embryos.

Plant cell culture

This technique is likely to continue to be keys of importance in the application of molecular techniques to crop improvement, where vegetative production is not possible, and techniques relevant to other molecular methods can be applied. These will enable breeders to try for new genetic combinations and the generation of genetic diversity will help to produce uniform and proven products. The introduction of valuable specific genes from distant and/or wild species into the species of interest either by a vector or by direct injection to widen the genetic base. Hence, the improvement of aromatic plants can be carried out by using both traditional and recently developed techniques.

Some aspects of **plant cell cultures** and **molecular techniques** are currently used and useful in the future for the production of aromatic compound, will be highlighted in the plant-biotechnology.

Plant biotechnology, in particular, **plant cell culture** is immensely important for the improvement of aromatic plants. Complete plants have already been regenerated from **callus culture**, **excised anthers**, **isolated protoplasts** of many aromatic plants. Many of the regenerated plants showed soma-clonal variation and selections were made for high alkaloid content cell lines. **Protoplast fusion** has also been achieved, but it is only involving members of Solanaceae family. Once a desired plant was regenerated, micro-propagation techniques were used to multiply and clone plant species. **Gene transfer** is possible from wild and related species to desired cultivars through **wide hybridization,** which includes **embryo rescue** systems. In many cases, accumulation of active compounds was found in **embryo tissue and embryo culture systems** providing "good harvests" of the compounds. **Plant suspension culture** is the most profitable system for **secondary metabolite synthesis**, but the techniques yields lower quantities of aromatic importance chemicals than the whole plant grown in the field. **Hairy-root cultures** are promising, but the cost is higher than leaf materials harvested from field crops.

Genetic transformation

It is another important area, which may provide dividends of increase and efficient *in-vitro* production of secondary metabolites in aromatic plants. The above techniques hold promise for the improvement of aromatic plants, but so far only a few experiments have been carried out for transformation using direct gene transfer methods in aromatic plants. It is hoped that research in this direction will be employed for the future improvement of aromatic plants.

Research work on different aspects of study are undergoing on different medicinal and aromatic plants at various institutes in India as given below.

Major breeding objectives of Aromatic Plants

In broad sense, the following objectives should be kept in mind as per need depending from crop to crop.

1. Collection and preservation of important aromatic plants in *in-situ/ex-situ* conditions.
2. Saving of endangered aromatic plants species from extinction.
3. Collection of native germplasm through explorations.
4. Diversification of agriculture by incorporating suitable aromatic plants of that region in traditional agriculture.
5. Conservation, documentation and sustainable use of important aromatic plants available of the region.
6. Genetic improvement/breeding for development of superior cultivars through various traditional approaches (introduction, selection of plants, mutation approach and clonal selection for vegetative grown plants); and modern plant breeding approaches (chromosome manipulation, biotechnology/molecular techniques *viz.* tissue culture, micro-propagation and production of secondary metabolites etc.) suitable to different crops as per feasibility and availability depending upon the economic plant parts (roots, stem, foliage, seeds or whole plant) used for their increased yield (herb, roots, rhizomes, inflorescence/flowers, seeds, resin/exudates etc.); and better quality of the produce for major quality parameters/chemical constituents.

State Agricultural & Horticultural Universities, ICAR and CSIR Institutes working on Aromatic Plants

S. No.	Name of the University/Institues	Abbreviation
1	Anand Agricultural University, Anand (Gujarat)	AAU
2	Assam Agricultural University, Jorhat (Assam)	AAU
3	Agricultural University Jodhpur (Rajasthan)	AUJ
4	Banda University of Agriculture & Technology,	BHAT
5	Bidhan Chandra Krishi Vishwavidyalaya, Kalyani (Bihar)	BCKU
6	Bihar Agricultural University, Islampur (Bihar)	BAU
7	Birsa Agricultural University, Ranchi (Chhatisgarh)	BAU
8	Chaudhary Charan Singh Haryana Agricultural University, Hisar (Haryana)	CCS HAU
9	Indira Gandhi Krishi Vishwavidyalaya, Raipur (Chhatisgarh)	IGKV

10	Jawahar Lal Nehru Krishi Vishwavidyalaya, Jabalpur (M.P.)	JNKVV
11	Kerala Agricultural University, Trichur (Kerala)	KAU
12	Maharana Pratap Horticultural University, Karnal	MHU
13	Maharana Pratap University of Agricultural & Technology, Udaipur (M.P.)	MPUAT
14	Mahatama Phule Krishi Vidyapeeth,, Rahuri (Maharashtra)	MPKV
15	Narendra University of Agricultural & Technology, Ayodhaya (U.P.)	NDUAT
16	Orissa University of Agricultural & Technology, Bhubaneshwer (Orissa)	OUAT
17	Dr. Punjabrao Deshmukh Krishi Vidyapeeth, Akola (Maharashtra)	PDKV
18	Rajmata Vijayraje Scindia Krishi Vishwavidyalaya, Mandsaur (M.P.)	RVSKVV
19	Tamil Nadu Agricultural University, Coimbatoor (T.N.)	TNAU
20	Uttar Banga Krishi Vishwavidyalaya, Kalingpom (W.B.)	UBKV
21	Uttarakhand University of Horticulture & Forestry, Bharsar (Uttarakhand.)	UUHF
22	Dr. Y. S. Parmar University of Horticulture & Forestry, Solan (H.P.)	YSPUHF
23	Dr. Y.S. Rajasekhara Reddy Horticultural University, Venkataramnnagudem (A.P.)	YSRHU
24	S. K. University of Agriculture Science & Technology, Kashmir (J&K)	SKUAT
ICAR & Central Institues		
25	ICAR - Directorate of Medicinal & Aromotic Plants and Betelvine, Boriavi-Anand (Gujarat)	DMAPR
26	ICAR – Indian Institute of Horticultire Research, Banglore (Karnataka)	IIHR
27	Dr. Rajendra Prasad Central Agricultural University, Pusa (Bihar)	DRCAU
28	Central Agricultural University, Pasighat (Bihar)	CAU
CSIR Institutes		
29	Central Institute of Medicinal & Aromatic Plants, Lucknow (U.P.)	CIMAP
30	National Botanical Research Institute, Lucknow	NBRI
31	Regional Research Laboratory, Jorhat (Assam)	RRL
32	Regional Research Laboratory, Jammu (J&K)	RRL
33	Institute of Himalaya Bioresource Technology, Palampur (H.P.)	IHBT
34	Central Institute of Medicinal & Aromatic Plants, Research Center, Banglore	CIMAP-RC
35	Central Institute of Medicinal & Aromatic Plants, Research Center, Hyderabad	CIMAP-RC

References

Chomchalow, N. 1980. Medicinal plants and spices in Asia. Pl. Genet. Resources Newsl., 44: 2-11.

Tetenyi, P. 1992b. Plant breeding and variety control of medicinal plants. In: Cultivation and Processing of Medicinal Plants (Ed.) Hornok., L. John Wile and Sons, NY, pp. 69-79.

Tetenyi, P. 1991. Biological preconditions for cultivation and processing of medicinal plants. In: The Medicinal Plant Industry (Ed.) Wijesekera, R.O.B. CRC Press, Boca Raton, pp. 33-41.

Sobti, S.N., Pushpangandan, P., Thapa, R.K., Aggarwal, A.G., Vashist, V.N., and Atal, C.K. 1978. Chemical and genetical investigations in essential oils of some Ocimum species, their F_1 hybrids and synthesized allopolyploids. J. Nat. Prod., 41: 50-55.

2

Plant Genetic Resources in Aromatic Plants

Introducton

The aromatic plants are the valuable source for flavours, fragrances and other pharmaceutical products. The aromatic plants possess odoriferous and volatile substances, which occur in the form of essential oils, gum exudate, balsam and oleo-resin. The chemical nature of these aromatic substances may be due to a variety of complex chemical compounds. Natural essential oils comprising a complex mixture of aroma chemicals, which are difficult to reconstitute in the chemical industry, fetch very high prices. India has a long perfumery tradition that dates back to over 5,000 years in Indus Valley Civilization. In the excavation of Harrappa and Mohanjodaro, a water distillation still and receiver have been recorded, whose shape resemble to the deg and bhabka currently used by attaries (traditional perfumers) of Kannauj in India (Gupta and Chadha, 1995).

Aromatic plants are very important part of agriculture and have been grown in India for their fragrance and therapeutic use from ages. Presently, several aromatic plants are in use for medication have come from our ancestors, inhabiting in different countries. The recent advances made in chemical and instrumentation technology led to the identification and isolation of principal chemical compounds diversely used in medicines, food, cosmetics, toiletries, bouquets, aerosols, paints, varnishes, insecticides etc (Arya *et al*., 2021).

However, the perfumery and allied industries consume much larger amounts of the natural fragrant material. Many of the aromatic plants are valued for naturally occurring essential oils and the bulk of the produce finds use in perfumery and the food flavouring industries. There are over 1200 aromatic plant species being used, out of which 40 species are traded throughout the world. Majority of raw materials of some of the indigenous species are still comes from the wild, not from the agriculture and plantation sector. As a result, the availability of the raw materials in the wild is declining at faster rate due to continuous exploitation by user agencies. The introduction of some of these into agriculture is at initial stage and need more efforts for cultivation. However, the diverse agro-climatic conditions prevalent in the country provide ideal habitat to grow these plants as substitute for diversification of agriculture.

World market potential v/s Indian share

The World market in the plant based drugs and essential oils is large and expanding vertically. Various kinds of essential oils traded in the international market with a large and consistent demand. As per the World market, at present for herbal health care products fetches more than 14 billion US $ as predicted by the world Bank, and it is estimated that it will further increase to 5 trillion US $ by 2050.

In global context, the Indian position in essential oil sector has increased from sixth to third in terms of volume of production (16-17%), and second in terms of the value (21-22%). Due to the concerted efforts of research institutes, universities, growers and industrial houses many crops could be taken on large-scale cultivation and many of them have made very good success. Mint is very live example of it, which made India as top producer. Many advanced countries, like Germany and Japan with very strong production base of synthetics also facing tough competition for price and quality with natural Menthol produced in India. Introduction of many new varieties in *Mentha* sp., Vetiver/khas, *Cymbopogon* sp., *Ocimum* sp., *Pelargonium* sp. etc., which gave higher yield coupled with higher percentage of desired quality components.

Evaluation and utilization of PGR

The Indian gene centre is one of the important regions of diversity of crop plants. It possesses over 17000 species of higher plants, of which over 40 per cent are endemic; and 3000-5000 species being of economic value. Like other plants, the genetic variability in aromatic plants in nature is due to prolonged changes in the habitat and environment. The natural gene pool still provides enormous variability for secondary metabolites and forms natural resources. In order to harness the potential of these natural resources, a holistic approach is required to regulate and manage these valuable resources in scientific manner.

In this context, collection, characterization, evaluation and utilization of these resources occupies a unique place in the field of PGR management. After agro-morphological and chemical evaluation, the potential of these resources can be harnessed for commercial production of raw material as well as to find out the newer source of aroma chemicals. We are confined to a few important aromatic crops, like, *Mentha sp.*, Vetiver, Palmarosa, Lemongrass, Basil, Geranium Rosemary etc., in which systematic work on augmentation, including germplasm explorations & collections, characterization and evaluation; including agro-morphological and chemical parameters and utilization in developing cultivars and varieties etc. has been undertaken in central and state universities in different parts of the country.

Mint/Pudina (*Mentha* spp.)

Genus *Mentha* belongs to family Lamiaceae with important species such as *Mentha arvensis*, *M. piperita*, *M. spicata* and *M. citrata*. It is mostly distributed in temperate and sub-temperate regions of world. It grows well in well drained fertile soils in tropical and sub-tropical climate. Mints are extensively cultivated in India, USA, Europe, south-east Asia and Latin American countries. The herb on distillation yield essential oil, which find use in pharmaceutical, food flavour, cosmetic, beverages and allied industries. Japanese mint (*Mentha arvensis*) oil represents the best source of menthol, used in pharmaceutical, flavouring and perfumery industry. India is the main producer of mint oil, menthol, dementholised oil, mint terpenes etc. Being the crop of exotic nature, the germplasm mainly augmented from different countries and evaluated for agro-morphological and chemical parameters. It has resulted into identification of promising accession, like EC-41911 having high herbage yield coupled with high oil content and it's quality.

Palmarosa (*Cymbopogon martinii* var. motia)

It belongs to family Poaceae and is native to India. It is grown on commercial scale in the states of Uttar Pradesh, Madhya Pradesh, Maharashtra, Andhra Pradesh and Karnataka. It is a drought tolerant plant and can also be grown under saline and alkaline soils conditions. Freshly harvested herb with leaves and inflorescence yield essential oil (rose like aroma), having high demand in perfumery, insecticidal and cosmetic industry. Its oil is used in perfumery, particularly, for flavouring tobacco and blending in soaps. Therefore, its oil fetches higher prices and has high position in international market. Presently, India is the main producer of its oil, along with other countries, including Brazil and Madagascar.

The crop is indigenous and the natural diversity is distributed in Vidarbha region of Maharashtra, southern hilly tract of Madhya Pradesh and Tarai region of Uttar Pradesh. Two crop specific explorations were undertaken by the scientists of NBPGR in diversity rich areas and collected 55 accessions of this important plant. These accessions were further evaluated for various agro-morphological traits and chemical constituents of oil (Sethi *et al.*, 1986). As a result, two promising accessions, namely, IW-31245 and IW-4498 were identified with high herbage yield coupled with high oil content.

Lemongrass (*Cymbopogon flexuosus*)

It also belongs to family Poaceae and native to India. Its natural diversity occurs in Western Ghats, comprising, Kerala, Karnataka and Tamil Nadu; and also in Sikkim and Arunachal Pradesh (Thomas, 1995). It is stemless and essential oil is obtained by distillation of its leaves. It is the richest source of citral, used as an industrial raw material for production of alpha- and beta-ionones and Vitamin A. India has been the main producer and exporter of its oil to the world market. The germplasm lines collected from the region were evaluated at KAU, Odakkalli centre resulted in the development of two varieties, namely, OD-19 and OD-440 for the irrigated conditions.

Vetiver/Khas (*Chrysopogon zizanioides*)

The generic name *Vetiveria* is a Tamil word meaning "the root is dug up" and species name "zizanioides" means "by the road side".. It belongs to family Poaceae and commonly known as khas. It is popular for its fragrant roots, providing valuable oil for perfumery industry. Naturally grown large populations are found to occur in wet and damp environments over marshy places in India. The plant is native to India and has been introduced in many tropical countries. It is distributed in seventy countries of Asian, African, Caribbean and pacific region along with Sri Lanka, Indonesia, Haiti, Jamaica, Zaire, Vietnam and Brazil for its sweet aromatic roots. Indian vetiver oil is highly priced and has export potential because of sweet, pleasant earthy woody balsamic note.

Explorations undertaken in diversity rich areas, like north eastern Rajasthan, especially Ghana Bird Sanctuary, Bharatpur, Sawai Madhopur and adjoining areas; and the central parts of Uttar Pradesh, especially Kannauj, Kanpur, Ghatampur, Musanagar etc. resulted in the collection of 94 diverse accessions. These accessions were later on characterized and evaluated for various agro-morphological and quality parameters of oil (Maheshwari, *et al.*, 1992). Further lines NC-66403, NC-66404 and NC-66416 were identified as promising for root and oil yield, including quality parameters producing earthy, woody, sweet, balsamic tape of oil. The oil of north Indian germplasm showed positive optical rotation, high vetiverol and has chemical maker khusinol, in addition to khusimol and khusol.

Basil/Tulsi (*Ocimum* sp.)

It belongs to family Lamiaceae and widely distributed in the tropical, sub-tropical and warmer part of the temperate regions. The maximum diversity of this species is known to occur in tropical rain forests of Africa. The leaves and other tender parts yield essential oil, used for various kinds of flavours in confectionery, baked goods, sauces etc. Its oil has various chemical compounds *viz.* pinene, camphene, mycrene, limonene, cis-ocimene, camphor, linalool, methyl chavicol, geraniol, methyl cinnamate and eugenol. European type of basil oil is considered to be of the highest quality because of its finest odour. The essential oil also possesses anti-microbial and insecticidal properties. These properties are due to the various constituents of essential oil, like methyl chavicol, eugenol, linalool, camphor, methyl cinnamate etc.

Systematic research work was done for its augmentation, characterization and evaluation for agro-morphological and chemical traits along with the identification of different chemotypes. Sixty accessions were augmented from indigenous as well as exotic sources and evaluated for agro-morphological and chemical parameters of oil composition. Based on oil composition, some of the exotic accessions were categorized into different chemotypes. These exotic accessions were rich in methyl chavicol (EC-338785 and EC-388890); rich in linalool (EC-174527, EC-388782, EC-388772 and EC-387838); and rich in methyl cinnamate (EC 338893 and EC 338893) Suneja *et al.*, 2005.

Geranium (*Pelargonium graveolens*)

It belongs to family Geraniaceae and is native to Cape province of South-Africa, chiefly grown in Spain, Morocco, Congo, Egypt, South France and East Africa. It was introduced in India on Shevaroy hills (Salem district of Tamil Nadu) in early period of the 19th century. It is presently grown commercially in Bangalore (Karnataka), Annamalai, Nigiris and Kodaikanal hills in Tamil Nadu. The leafy twigs are main source of its oil having high perfumery value and widely used in soap, perfumery and cosmetic industries. There are several types of geranium oils, but the best one is considered to be of Reunion origin. This oil has a heavy rose and minty odour as compared to delicate odour of Algerian type. Different types of geranium germplasm have been augmented and evaluated at NBPGR Regional Station, Bhowali. An accession IC-553286 was identified as high citronellol content and registered as unique germplasm by the Plant Germplasm Registration Committee and accepted by the industry for its lemon like odour.

References

Arya RK, P. Kumar,GS Dahiya, JM Sutaliya, R Kumar, AK Chhabra 2021. Medicinal Garden at a Galance. MAP Section (GPB), CCS HAU, Hisar. Pp1-42.

Gupta R., and K.L. Chadha. 1995. Medicinal and Aromatic Plants Research in India. In: Advances in Horticulture, Vol. 11, Medicinal and Aromatic Plants Eds. K.L. Chadha and Rajendra Gupta. Malhotra Publishing House, New Delhi, India. pp. 1-45.

Kumar Ashok, S.K. Pareek, Archana P. Raina and S.K. Mishra. 2008. In: Proceedings of National Seminar on "Recent Trends in Research on Spices and Aromatic Plants" held at CCS HAU, Hisar on Sept., 8-10, 2008, pp. 22-25.

Maheshwari, M.L., V.K. Srivastava, K.L. Sethi, and R. Gupta. 1986. Vetiver collection from Bharatpur, natural variability in physic-chemical constituents. Indian Perfum., 30 (4): 471-78.

Sethi, K.L., M.L. Maheshwari and R. Gupta. 1986. Genetic resources of palmarosa oil grass and their utilization. Indian Perfum., 30(2-3): 309-13.

Suneja, P., J. Mohan, Ashok Kumar, D.S. Pilania and S. K. Pareek. 2005. Quality evaluation of some exotic lines of Ocimum germplasm. Indian J. Plant Genet. Resour., 18 (1): 124-25.

Raina, Archana P., Suresh Walia, K.K. Sharma, Z. Abraham and S.K. Mishra. 2007. Component of essential oil of greater galangal (Alpinia galanga) rhizomes from Kerala, South India. Indian Perfum., 51(2): 54-58.

Thomas J. 1995. In: Advances in Horticulture Vol. 11 – Medicinal and Aromatic Plants. Eds. K.L. Chadha and Rajendra Gupta, Malhotra Publishing House, New Delhi, India, pp. 717-733.

3

Significance of Agro-biodiversity in Aromatic Plants

India is endowed with rich treasure of aromatic plants of great industrial importance in its forests, wetlands and marine areas. India is among the 12 mega biodiversity centers of the world, which accounts for about 8 percent of the total biodiversity recorded so far. Despite of having only 2.4 percent land area of the world, India has been divided into 16 Agro-climatic zones having about 49000 plant species. Out of which 15000 species belong to medicinal and aromatic plants group. Out of these, 1500 medicinal and aromatic plants have identified, but only 500 species are commonly used in the preparation of ISM&H drugs. Besides, more than 150 plant species have been found endangered and are at the verge of extinction.

Indian export of raw drugs based on medicinal and aromatic plants has grown steadily at 25 % over the last few years. WHO has forecasted that the global market for herbal products is expected to be around US$ 5 trillion by 2050, and further emphasized the need for revitalization, and expansion of traditional system of medicines. But merely 150 medicinal and aromatic plants species are being reported to be used commercially on fairly large scale (Chandel *et al.*,1996).

The global market in plant based pharmaceuticals in growing rapidly day-by-day leading to a sizeable increase in demand for herbal plants. Even today, 80-90% of naturally growing wild populations have remained the main source of raw materials used for drugs in India without applying scientific management techniques essential to sustain the growth and preserve useful properties essential to determine the efficacy of the medicines. The wild source is now being depleted and exploited without scientific harvesting techniques/ practices. The illegal export of several rare species has shot up alarmingly due to laxity in enforcement of rules on trade in wild flora, especially medicinal and aromatic plants. This has led to considerable genetic erosion and loss of biodiversity pertaining to medicinal and aromatic plants of great value.

Inspite of the fact that over 75% of the world population still relies on plants and plant extracts for their healthcare, more than 30% of the entire plant species have been in use for medical, aromatherapy, cosmetic & perfumery and bio-pesticide industries. The heritage of aromatic plants use in India has an ancient history and known for its repository in their natural habitats. It is revealed that about 8000 herbal remedies have been codified in Ayurveda, and 1270 species have been reported in preparation of various drugs. In developed countries, plant based drugs constitute around 25% of total drugs, whereas in developing countries its contribution is upto 80 percent. Therefore, the importance of herbal plants in Indian context cannot be overlooked.

India- treasure of plant biodiversity

India has very diverse agro-climatic and regional topography. The preponderance of indigenous tribal populations and their ethnic groups have contributed immensely for its diversification over a period of time. Out of 25 hot spots of biodiversity in the world, two mega diversity hot spots located in India.

Biodiversity status at national level

S. No.	Taxonomic group	No. of species
1	No. of plant species	45000-49000
2	No. of higher plants species	18000
3	No. of Ethano-botanical plants	9500
4	No. of Ethano-medicinal plants	7500
5	No. of medicinal plants	3500
6	Plants species used in Ayurveda	2351
7	Plants species used in Siddha	1121
8	Plants species used in Homeopathy	482
9	Plants species used in Unani	880
10	Plants species used in Amchi + Folk	600 + 900
11	Plants species used in Allopathy (Modern medicine)	30-80
12	Commercial medicinal plants	150
13	Aromatic plants species	1300
14	Aroma industries exploited only	42

****Source:*** Kumar and Asija, 2004.

Phyto-geographical regions of the country

Indian forests have been divided into eight (8) phyto-geographical regions, which are rich homes of aromatic plants as given below.

Region	Plants species diversity
Western Himalayas	*Mentha sp., Ocimum sp.,*
Eastern Himalayas	*Mentha sp.,*
North Eastern Region	*Curcuma spp., Cympogon spp.,*
Gangetic Plains	*Curcuma spp., Vetiveria zizanioides*
Semi-Arid Region	*Cympogon spp*
Western Ghats	*Curcuma spp.,*
Eastern Ghats	*Curcuma spp.,*

Source: Gupta *et al.*, 1997.

Conservation techniques of Plant Bio-diversity

Government of India (1972) made an Indian Forest Act, to Impose ban of the export of wild growing certain plants of medical and aroma industry use own account of over-exploitation. Conservation and protection of these plants needs to be adopted appropriately on high priority by the following two basic approaches of conservation.

A. *In-situ* conservation

It relates to the conservation of aromatic plants in their natural habitats, and thus retaining their full natural evolutionary potential. Some of the protected areas known as biosphere reserves, national parks, wild life sanctuaries and sacred groves have been established for conservation of plants with great economic values.

i. **Biosphere reserve:** An area of about 4.2 percent has been reserved for this purpose. Presently14 biospheres have been established in the country to carry out the major activities as given below.

a) **Setting up of aromatic plants conservation programme in Asia:** Such programmes are launched for dealing with specific key research issues, networking and collaboration in South East Asia region for sustainable and equitably use if MAPs in the region. Such projects are funded by IDRC and Ford Foundation, New Delhi.

b) **Characterization of Biodiversity and conservation of reserves of MAPs:** CSIR-CIMAPs, Lucknow (U.P.) is playing major role in monitoring, guiding and co-ordination of various projects with a vision of innovation and empowerment in science and business related MAPS.

ii. **National Parks and Wild Life Sanctuaries**: The protected areas encompassing the biological diversity in its natural condition comprised of 89 national parks and 448 wild life sanctuaries with an about 4.5 percent area of the total geographical area in the country.

B. *Ex-situ* Conservation

The cultivation and maintenance of MAPs outside its natural habitats either in botanical gardens or other suitable sites through long term preservation of plant propagates in gene banks and in plant tissue culture repositories.

i) **National Gene Banks:** CSIR and DBT, New Delhi established three national gene banks at CIMAP, Lucknow; TBGRI, Thiruvananthapuram; and NBPGR, New Delhi with facilities of a field bank, seed bank, in-vitro bank and cryo-bank.

ii) **Botanical Gardens**: Presently, a network of 344 botanical gardens spread over different states supported by Government as well as Private sector are playing a pivotal role in *ex-situ* conservation, especially those facing imminent threat to extinction.

Government of India (GOI) has now set up a Central Nodal Agency known as Medicinal Plants Board (MPB) during November, 2000 under the Ministry of Health and Family Welfare, New Delhi for the development and sustainable use of MAPs. This will supervise and co-ordinate policies and strategies for conservation, harvesting, cost-effective cultivation, R&D, processing and marketing of raw materials for development of this sector.

Crop improvement in aromatic plants is dependent on plant genetic diversity with wider gene pool for natural evolutionary adjustment according to changing environment, biotic and abiotic conditions. Rich plant wealth may be used directly for cultivation or as a source for transfer of useful traits for breeding improved varieties. Therefore, collection, conservation and sustainable use of these genetic resources and their wild relatives require specific attention keeping in view their overall significance for day to day needs in human life in agriculture and food security.

Future thrust and developmental needs

- Areas of severe genetic erosion should be identified.
- Deforested land should be re-forested with main focus on APs.
- Identification of germplasm for more potent and newer drugs compound and aroma-chemicals.
- Quality testing, developing standards and use of accurate analytical methods.

- MAPs should be introduced as inter-crop, cash crop, border crop and under cover crops in partial shades.
- There is a need for indigenous traditional knowledge practices (ITK) at national level.
- Urgent need to update the Red Data Book of Indian APs.
- Conducting large scale training and awareness camps on different aspects of commercialization of APs.

Registration of Plant Germplasm

National Council of Agricultural Research (NCAR) considered registration of germplasm as a need of the hour, and due recognition to the scientists who developed improved cultivars for the said purpose. Similarly, the developer of released varieties, scientists associated with the development of improved germplasm and genetic stocks for new sources of resistance, male sterility, mutants, cytogenetic stocks should be recognized in order to preserve most of such valuable materials. Now with the recent developments concerning IPR and related issues, due recognition of these materials have become more important.

In view of this, NCAR constituted a committee to suggest guidelines for identification & registration of value-added germplasm. The recommendations of the committee have been approved by the Council for consideration. **National Bureau of Plant Genetic Resources (NBPGR),** New Delhi has been identified as the Nodal Agency for implementation. It is now hoped that this registration of germplasm will encourage better sharing of genetic material among the scientists.

References

Bhardwaj, K.R. and Rajesh Gulia. 2012. An appraisal of present status of biodiversity management for medicinal and aromatic plants. In proceedings of seminar on "New perspectives in Aromatic and Medicinal Plants." held at CCS HAU, Hisar on Feb., 8-9, 2012, p. 9-17.

Chandel, K.P.S., G. Shukla and N. Sharma. 1996. Biodiversity in Medicinal and Aromatic plants in India: Conservation and utilization, p.1-239, NBPGR, New Delhi.

Gautam, P.L. 1997. Agro-biodiversity in Indian gene centre conservation strategies. Key note address in workshop on "National concern for Management Conservation and Use of Agro-biodiversity" held at Shimla, Oct. 15-16, 1997.

Gautam, P.L. and S. Kochhar. 1997. Status of international undertaking on plant genetic resources IUPGR) of FAO and further issues. Workshop on "National concern for Management Conservation and Use of Agro-biodiversity" held at Shimla, Oct. 15-16, 1997, Souvenir p. 29-33.

Gupta, Veena, B.M. Singh and P.L. Gautam. 1997. Agro-biodiversity Conservation of Medicinal and Aromatic Plants in India. Workshop on "National concern for Management, Conservation and use of agro-biodiversity", Souvenir, p. 46-49.

Gupta, R. and K.L. Chadha. 1995. Medicinal and Aromatic Plants in India. In "Advances in Horticulture", Vol. 11, Eds. K.L. Chadha and Rajendra Gupta, Malhotra Publishing House, New Delhi, p. 1-44.

Kumar, U. and M.J. Asija. 2004. Biodiversity: Principles and conservation. Agribios India. Jodhpur, pp. 1-234.

4

Role of Biotechnology in Aromatic Plants

Introduction

There has been increasing demand of plant based raw materials for manufacturing pharmaceuticals, flavours, fragrances, cosmetics & other related products. Among these, ginsenosides, taxol, artemisinin etc. are some of the biomolecules used in making drugs with great demand in the market. Aromatic plants are an integral component of research and development in the pharmaceutical and aroma industry.

In India, the annual growth in this sector is around 15-20 percent. There are approximately 7500 plant species of medicinal value, including aromatic plants having significant scope for phyto-pharmaceuticals and cosmetic & perfumery industry. Because raw materials are still harvested from wild growing populations in the forests, introduction of these plants in agriculture is need of the hour for further research efforts to bring them into commercial cultivation as these plants also promise large benefits to farmers, if brought under cultivation. This can be achieved by screening germplasm from natural flora to select suitable genotypes for sustainable use in the prevalent cropping system.

Plant tissue culture technically known as **micro-propagation** and broadly defined as a collection of methods used to grow large number of plants *in vitro* (in an aseptic and closely controlled environment). This technique is effective because almost all plant cells are totipotent, meaning thereby that each cell possesses the genetic information and cellular machinery to generate an entire organism. Micro-propagation, therefore, can be used to produce large number of plants, which are genetically identical to parent plants as well as to one another (Raveen *et al.*, 1999).

Strategies for micro-propagation

As standard protocol it requires sterile environment in laboratories for medium used to grow the plant tissue and the environment surrounding the tissue culture. The presence of any bacterial, fungal, algal or viral contaminants could potentially rob the desired nutrients of plants provided by the culture medium and have devastating effect on their growth (Kyle, 2002).

Then tissue can be collected from the leaf, shoot, and bud or root, because each of these cells are totipotent. The tissue sample is then placed on an aseptic, nutrients-rich medium, where its cells begin to grow and develop into the desired plant product (Lineberger, 2002). The nature of the medium and nutrients contained depend upon the type of plant being grown and the properties that the grower wishes to express. Finally, the developing tissue should be maintained in a closely controlled chemical and physical environment (green house) to achieve the best results.

Generally, MS basal (Murashige and Skoog, 1962) medium with B_5 vitamins is used for culturing, supplemented with different growth regulators and additives. These are either auxins or cytokinines or combination of both with gibberellins or additives (adenine sulphate, glutamine, casein hydrolysate etc.). After culturing of the cultured vessel containing explants with medium should be kept under 25 ± 2^0C and under 16/8 hrs. photoperiod. Most commonly used explants are leaves, meristem, shoot tips, stem and roots (vegetative parts). Besides, reproductive parts *viz.*, pollens, embryo, seeds and spores can also be used for culturing (Jain and Thakur, 2005).

Micro-propagation can be done by two pathways

i. Direct regeneration

ii. Indirect by somatic embryogenesis or by callus formation.

There are certain intermediary stages in micro-propagation from explant to plant, such as establishment of *in vitro* culture, shoot multiplication, root development, acclimatization and transfer of plantlets to soil.

Generally, in shoot multiplication, the medium must be of variable concentration of cytokinines from 1-2 mg/l up to 30 mg/l used alone or with combination of auxins (commonly 0.1-1.0 mg/l). Higher concentration of cytokinines induces adventitious buds and retard shoot growth.

In rooting, low salt i.e. ½ or ¼ MS basal medium should be used with around 0.1-1.0 mg/l of auxins. During the process of acclimatization tissue culture medium to pot and ultimately to the field, high humid environment is required.

There are many factors which influence micro-propagation, such as genotype, nature of the explant and composition of medium as the main factors; but the initial response of the culture tissues may vary with overall physiological status of the donor plant, season in which explants taken and the environment where they are kept after culturing.

Acclimatization of regenerated tissue is the final, but necessary step in all plant micro-propagation scheme where plant has to adapt to lower relative humidity, higher light intensity, fluctuating temperatures and constant biotic stress. Excessive water loss and impaired photosynthetic machinery are the two major problems.

Need for micro-propagation

Micro-propagation has special significance, when only small amount of tissue or only few plants are available; and large numbers of plants are to be obtained within a short period. Micro-propagation technique plays an important role in conservation and management of botanical collections. It is also quite helpful technique in solving out the problems of breeders and farmers on commercial scale. It is used where generally there is problem in seed germination, seed viability, seed setting, slow propagating plants, pollen or ovular sterility, germplasm conservation, obtaining virus free plants, introducing new cultivars and also in getting quality planting material on commercial scale (Karp and Jones, 1985).

Vegetatively propagated plants frequently become infected with disease caused by different viruses. It has been observed that apical meristem may have not or low chances of presence of virus particles. So by culturing apical meristem provided with heat and chemical treatment, we could get virus free plants. Around 65 virus free species have been generated till date. There is problem of viability, when we store seed for germplasm, but this can be overcome if germplasm is stored in form of shoot cultures at low temperature, say 8^0C in maximal medium and also an alternative method is cryo-preservation, which is done in liquid N_2 at -192^0C. In the sexual crosses, where parents are taxonomically distant, barriers that prevent successful hybrid production, can be achieved by embryo rescue technique of micro-propagation, where immature embryos can be excised and cultured after that hybrid plants can be obtained. Production of haploids and dihaploids by anther and ovary culture can also be obtained by micro-propagation.

Aromatic plants listed under are cropped on commercial scale and farmers are being provided planting materials by tissue culturing technique. *Curcuma domestica, Curcuma longa, Mentha piperita, Mentha spicata, Cymbopogon citratus/Ocimum sanctum, Ocimum basilicum, Ocimum canum* and *Ocimum grattisimum,*

Each of these plants have one or another problem such as fruit drop, setting of seed and some do not have commercial approach of planting material in field. These problems can be overcome by micro-propagation. Some of these plants can be cultivated on wastelands, created by deforestation, erosion or neglected areas.

Performance of tissue cultured raised plants is quite good as the survival rate in the field condition is around 80-90 percent. It is also possible to make plants varied in producing secondary metabolites by using chemicals.

Production of secondary metabolites

Aromatic plants contain many secondary metabolites e.g. flavonoids, saponines, lignins, etc., which play an active role in ISM and allopathy. The commercially used compounds are often those that are difficult to synthesize. Synthetic methods often those that are difficult to synthesize. Synthetic methods often produce unwanted stereo-isomer and contain impurities from side reaction and traces of reactants. The plant ability to produce secondary metabolites in considerable quantities can be used by growing aromatic plants as crop, harvesting the relevant plant parts and extracting the compounds from these parts. Alternatively, these metabolites can be produced by culturing plants cell *in vitro*. Using this approach, the secondary metabolites can be obtained under controlled conditions, unaffected by seasonal variation in a short period of time. Plant tissue culture can also give valuable insight for understanding their biosynthetic pathways and engineering the pathways for over production of targeted metabolite in a model plant system.

Hairy root culture

The secondary metabolites can also be produced *in-vitro* by hairy root cultures. *Agrobacterium rhizogenes* is a gram negative soil microbe, which infects plants cell by transferring its T-DNA present in Ri plasmid in plant cells. The stable integration T-DNA in plant genome takes place, and this causes induction of hairy roots from the site of inoculation, which may be established in medium. Three genes rolA, rolB and rolC present in the T-DNA of Ri plasmid account for the hairy root formation. These cultures show high growth coupled with genetic stability and production of secondary metabolites. Hairy roots exhibit auxin autonomy which leads to their phytohormone independent proliferation in simple defined media generating root clones. Genetic stability of hairy roots facilitates the process of designing and scaling up the process at the commercial level. Root cultures can be a potential alternative system for production of pharmaceutical proteins also. Some of these phytochemicals have been scaled up 2000L to 20000L in bioreactors. Different types of bioreactors can be employed, which can give adequate mixing for proper aeration and efficient transfer of nutrients to plant cells with lowest power input and cell shearing.

Constraints

Plants tissue culture has become a boom for the production of many important aromatic plants and phytochemicals as the list of successfully micro-propagated plants are enlarging day by day, but this technique also has its own limitations and problems listed as under:

i. Lack of expression of totipotency in many plants.
ii. Genetic instability due to somaclonal variation in micro-propagated plants.
iii. Contamination is a major problem for precautions but no cure.
iv. In woody plant species, it is difficult to initiate roots.
v. Vitrification is also a major problem.

Future prospects

Micro-propagation is the technique in spot light and considerable scope, still exist for improving basic protocols for recalcitrant in particular. The automated method of micro-propagation with the involvement of robotics and computers can help in reduction in the cost of production and help to extend the technology to a wide variety of plants for large scale production of micro-propagules (Jha, 2005). However, it mean that tissue culture based technologies could replace the traditional plant breeding. It is going to remain a broad spectrum technology under pinning many disciplines, from propagation and conservation to genetic manipulation and also embracing biotechnological strategies for the screening and marketing of pharmaceutical and agro-chemical products.

References

Flores H.V. and Medina-Boliver F. (1995). Root culture and plant natural products: "unearthing" the hidden half of plant metabolism. Plant tissue culture and Biotechnology, 1(2): 59-74.

Hazarika B.N. (2003). Acclimatization of tissue cultured plants. Current Science, 85: 1704-1712.

Jain D.K. and Thakur T. (2005). Vegetative nursery plant production through micro-propagation. Agrobios newsletter. Pp. 14-16.

Jha V.B. (2005). Micro-propagation as an enterprise. Agrobios Newsletter. Pp. 32-33.

Jones M.G.K. and Karp A. (1985). Plant tissue culture technology and crop improvement. Advances in biotechnological processes, (5) pp. 91-121.

Lowe K.C., Davey M.R. and Power J.B. (1996). Plant tissue culture: past, present and future. Plant tissue culture and Biotechnology. 2(4): 174-141.

Murashige T. and Skoog F. 1962. A revised medium for rapid growth and bioassays with tobacco tissue cultures. Physiol. Plant 15: 473-497.

Raveeen Peter H., Ray F., and Eichorn Susan E. (1999) Biology of plants. 6th ed. (New York: W.H. Freeman and company).

Thorpe T.A. (1981). Plant tissue culture: methods and applications in agriculture. (Academic press, New York).

Van Huylenbroeck J.M. and Debergh P.C. (1996). Physiological aspects in acclimatization of micro-propagated plantlets. Plant tissue culture and Biotechnology. 2(3): 136-141.

Yadav, R. C., Vandanna Saroha, Anuya Chaudhary and Neelam R. Yadav. 2005. Published in Course Compendium of Winter School sponsored by ICAR, New Delhi on "Advances in Medicinal, Aromatic and Under-utilized Plants Research" held at CCS HAU, Hisar from Sept. 29 to Oct., 19, pp. 110-13.

5

Significance of Essential Oil Quality in Aromatic Plants

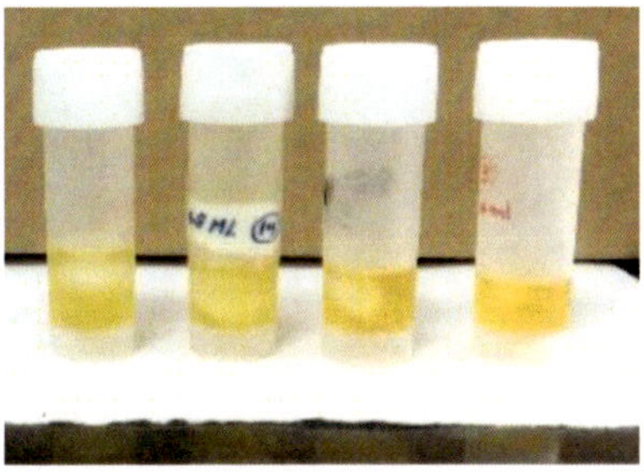

The separation of essential oils responsible for odour/fine smell from different plant-parts by gentle heating is known by centuries ago. A large number of such components being volatile are called as 'Essential oils'. Presently, it also covers many of the less volatile constituents. The essential oils (extracts of the plant) with different types of flavours/odours are aromatic components, which can be separated by steam/hydro-distillation or extraction by solvents.

Essential oils largely contain terpenic compounds, made up of two or more isoprenyl units (C-5). Besides, a large number of other compounds which have nothing to do with isoprene units, have also been reported by various workers. Essential oils are of quite use in perfumes, flavour and aroma-therapy industries and their economic production is very much essential from farmers' point of view. The optimum stage of harvest is very crucial for high quality and quantity of the essential oil from different parts of the plants (leaves, inflorescence, entire herb, roots, seeds, rhizomes etc.).

The quality of the essential oils largely depends on the kinds of the metals (copper, aluminium, mild steel, stainless steel) used in distillation unit, as some of the essential oils react with free acids present in the oil and thus imparting off-smell of aroma and colouration. Presently, distillation units made up of stainless steel are preferred which has no reaction to phenolic, acids, etc., and oils produced are very clean, less coloured and of high quality preferred by the cosmetic and perfumery industries engaged in formulation of various types of essence and scents.

Besides, storage of essential oils after distillation is of immense significance as these are made up of the terpenoids, phenolics, acids, sulphur compounds etc., likely to go under oxidation, polymerization etc. Therefore, for increasing shelf-life of the essential oils, it should be subjected to removal of off-odours moisture and suspended particles before storage in glass, stainless steel, aluminium wares, galvanised or lacquered drums. The containers should be full up-to brim and kept in cool and dark place to prevent from oxidation, polymerization and decomposition.

Finally, the essential oils thus preserved must to analysed for its quality

i. Determination of physico-chemical constants,

ii. Analysis by instrumental methods for its major chemical constituents responsible for high quality, and

iii. Performance of organoleptic tests in standard laboratories set for the said purpose before sale to have reasonable price in the market.

Keeping this in view, the major chemical constituents responsible for high quality of important aromatic plants are given below:

Common name	**Scientific name**	**Economic parts used**	**Major chemical constituents**
Palmarosa/ Roshagrass	*Cymbopogon martini var. motia*	Fresh herb	Geraniol (>80%)
Lemongrass	*Cymbopogon flexuosus*	Fresh herb	Citral (>75%)
Citronella grass	*Cymbopogon winterianus*	Fresh herb	Citronellal (>35%) Citronellol (18-20%) Geraniol (25%)
Mint/Pudina	*Mentha arvensis* (Japanese mint)	Fresh herb	Menthone (25-35%) Menthol (30-40%)
Basil/Tulsi	*Ocimum basilicum*	Fresh herb	Methyl chavicol (>70%) Linalool (>15%)
Khas/Vetiver	*Chrysopogon zizanioides*	Fresh roots	Vetiverol (45-70%)
Patchouli	*Pogostemon patchouli* Pallet.	Dry leaves	Patchouli alcohol (35 40%)
Geranium	*Pelargonium graveolens L.*	Fresh herb	Citronellol (>20%) Geraniol (15-20%) Linalool (10-15%)
Anise	*Pimpinella anisum L.*	Seed	Anethole (> 90%)
Celery	*Apium graveolens L.*	Leaves Stalk Seeds	Carbohydrates (5-10%) Carbohydrates (<5%) B-Selinene (10-15%)
Dill seed/Sowa	*Anethum graveolens* L.	Seeds	D-Limonene (20%) Carvone (20%)
Chamomile	*Matricaria chamomilla L.*	Flowers	Chamazulene (upto 15%)
Jasmine	*Jasminum grandiflorum L.*	Flowers	Indole (2-3%)
Rose	*Rosa damascena Mill.*	Fresh flowers	Geraniol (60-70%) Citronellal (35-40%)
Davana	*Artimesia annua/pallens*	Flowers	Esters (65%) Hydrocarbons (20%)

6

Extraction Methods of Essential Oils in Aromatic Plants

Introduction

The quality of essential oils is very complex. An individual oil may have hundreds of ingredients, the principle components being a group of complex substances known as terpenes and their compounds or derivatives. Due to this reason a single essence has a wide range of therapeutic actions.

The chemical constituents of plants are of two kinds: products of primary metabolism and products of secondary metabolism. The first are mainly carbohydrates, amino acids and fixed oils produced by light-absorbing (photosynthetic) processes. The latter group of chemicals arises from the primary metabolites. They include glycosides, terpenoids, alkaloid and essential oils. Essential oils lack the water soluble constituents such as tannins, bitter compounds sugar, mucilages and pectins, playing important role in the medical action of the plant.

The essential oils have been credited with possessing almost all therapeutic properties in a concentrated form. The therapeutic properties of the whole oil are that result of synergism: an interaction of all its chemical constituents working harmoniously together so that the whole essential oil becomes more potent than the sum of its individual parts.

The essential oil lacks certain chemical compounds found in the raw plant material. It is still a highly complex substance, that it cannot be replicated in the laboratory. Even though high-tech processes such as gas liquid chromatography (GLC) can separate out main constituents of essential oils. By looking at the chromatogram, it is impossible to isolate the numerous trace elements including number of unidentified compounds which make up the whole. For this reason, synthetic or nature identical aromatic oils never quite the same as organic essences. The essential oils from aromatic plants are for the most part volatile and thus, lend themselves to several methods of extraction such as hydro-distillation, water and steam distillation, steam distillation, and solvent extraction (ASTA 1968, Guenther 1972, Heath 1981).

The specific extraction method employed is dependent upon the plant material to be distilled and the desired end product. The essential oils which impart the distinctive aromas are complex mixtures of organic constituents, some of which being less stable, may undergo chemical alterations when subjected to high temperatures. IN this case, organic solvent extraction is required to ensure no decomposition or changes have occurred which may alter the aroma and fragrance of the end product. Newer methods of essential oil extraction such as using supercritical CO_2 which yield very high quality oils are commercially used, but are less common and beyond the financial means of most processors. The recovery of non-volatile essential oils is also obtained by solvent extraction although the process is more difficult and complex than the recovery of the volatiles. This process yields an aromatic resinous product known as an oleoresin, which is more concentrated than an essential oil and which has wide application in the food industry (Heath 1981).

1. Essential oil extraction methods

Various methods are used for the extraction of essential oils. This includes distillation, cold pressing and in a few cases, non-toxic solvent extraction. A good therapeutic grade essential oil is produced by using non-toxic procedures that preserve the life giving vital essence of the plant. Essential oils vary greatly in quality and not all are therapeutic grade. Three main methods i.e. distillation, expression and solvent extraction are discussed in brief here:

a) **Distillation:** Distillation is still the most economical method of extracting essential oil from spices and aromatic plant materials. The main advantage of distillation is that it can generally be carried out with some very simple equipment, close to the location of plant production. Even in relatively remote locations large quantities of material can be processed in a relatively short time. Distillation is less laborious and required lower labour skill than solvent extraction. Adopting the simplest or cheapest extraction method, however, may prove to be false economy because of low yield, poor or highly variable oil quality and low market value. Some of the distillation methods are: hydro/water-distillation, water and stream distillation, steam distillation and fractional distillation.

i. **Hydro-distillation:** It is the simplest of the distillation methods. The plant material is mixed directly with water in a still pot. A perforated grid may be inserted above the base of the still pot to prevent the plants material settling on the bottom and coming in direct contact with the heated base of the still and charring. Although this method is simplest and cheapest, but the quality of the oil has the greatest potential to be modified due to the effects of direct heating and the water contact.

ii. **Water and steam distillation:** The basic still design is very similar to that of hydro distillation. The plant material is packed into the still pot fitted with a grill or perforated plate above the boiling water. The capacity of the still pot volume is reduced but it may be possible to achieve a high packing density because the plant material is not suspended in the water.

iii. **Steam distillation:** It is the process of distilling plant material with the steam generated outside the still in a boiler. As in the water and steam distillation system the plant material is supported on a perforated grid above the steam inlet. Wherever possible, the still vessel condenser and separator should be fabricated from stainless steel. Skills for maintenance and repair of equipment/machinery should be of primary consideration in the design of the distillation system. Seeking professional advice is also critical.

iv. **Fractional distillation:** It refers to a normal distillation process, but instead of the essential oil being collected continuously, it is collected in batches (the fractions that are referred to) and material normally so extracted is Ylang-Ylang.

b) **Expression:** Expression is a cold pressed method of extraction, which is mostly used in the extraction of citrus essential oils. Some of the expression methods are:

i. **Sponge extraction**: Most citrus essences are extracted by means of Sponge extraction expression. The fruit pulp is removed with the rind and pith then soaked in warm water to make the rind more pliable, since the pith of the fruit absorbed the water. After the fruit has absorbed the water and become more elastic, it is inverted which helped to rupture the oil cells and a sponge placed next to the rind. It is then squeezed to release the volatile oil, which is then collected directly into the sponge. As soon as the sponge become saturated with oil, it is squeezed and the essential oil is collected in a vessel and then decanted.

ii. **Machine abrasion:** A machine strips off the outer peel of citrus like fruits, which is then removed by running water and is then fed into a centrifugal separator. The centrifugal separation is done extremely fast but it should be noted that due to the fact that the essential oil is combined with other cell content for some time, some alteration could occur due to enzymatic action.

c) **Solvent Extraction:** The broad term of solvent extraction does not only refer to chemical solvents like hexane, but also to other forms – such as solid oil and fat as well as carbon dioxide. Solvent extraction is particularly suitable for botanical material that has a very low yield of essential oil, or where it is made up of mostly resinous components and as such delivers a far finer fragrance than that of distillation. During this type of extraction, non-volatile components of the botanical material – such as waxes and pigments are also extracted and in some cases this is then removed during another process.

Some of the solvent extraction methods are as below

i. **Maceration method**: The flowers are soaked in hot oil to have their cell membranes ruptured and the hot oil then absorbs the essence. The oil is then cleared of the botanical material and decanted. This is very such the same technique used in solvent extractions, where solvents are used instead of the hot oil.

ii. **Solvent extraction method**: The essential oils are extracted by using solvents such as petroleum ether, methanol, ethanol or hexane and is often used on fragile material such as jasmine, hyacinth, narcissus and tuberose, which would not be able to handle the heat of steam distillation.

A solvent extracted essential oil is very concentrated and close to the natural fragrance of the material used. Although solvent extraction is used extensively, but it is believed that it should not be used for aromatherapy oils, since a residue of solvent could be present in the finished product. Some reports site a solvent residue of 6-20 percent still present in the finished extraction, but this was normally the case when benzene was the standard solvent used.

With hexane (a hydrocarbon) as the solvent material, the solvent residue goes down to about 10 ppm (parts per million) and this is an extremely low concentration of solvent in the resultant production. As mentioned, benzene is no longer used in the extraction method, since it is regarded as carcinogenic (cancer forming). After the plant material has been treated with the solvent, it produces a waxy aromatic compound referred to as a "concrete".

iii. **Hyper-critical carbon-dioxide extraction (CO_2) method:** It is a fairly new way to extract essential oils from botanical material and although a bit on the expensive side, does yield good quality oils. Carbon dioxide becomes hypercritical at 33 degrees Celsius, which is a state in which it is not really gas or liquid, but has qualities of both, and is an excellent solvent to use in the extracted of essential oils, since the low temperature required and the fact that the process is near to instantaneous. The carbon dioxide is furthermore insert and therefore does not chemically interact with the essence that is being extracted. To remove the carbon dioxide solvent, there is a need to remove the pressure under which it is kept. This process has to take place in a closed chamber as the hypercritical pressure required for carbon dioxide is 200 atmospheres – that is 200 times the pressure of normal atmosphere. To achieve this type of pressure some heavy-duty stainless steel equipment is required and therefore high capital investment is needed for this extraction method.

2. Quality of essentials oils

Despite the continuing invention of synthetic chemical, essential oils/extracts of aromatic plants still remain the most important part of the fine fragrances. Most of the components of the essential oils are generally referred as safe in comparison to synthetic chemicals. Many essential oils are now-a-days used in perfumery, cosmetic, flavour and pharmaceutical industries. Therefore, the knowledge of quality of essential oils and their isolates is most important. The quality of essential oils of important aromatic plants is discussed below:

i. **Lemongrass (*Cymbopogon flexuosus*)**: The oil of these species is the most important source of citral which is starting material for the preparation of ionone. The oil is also used in low cost perfumes, soaps and other laundry products. IN a good quality lemongrass oil (Guenther, 1977), contents of aldehydes i.e. citral should be maximum (70-85 %). Citral is found in two isomeric forms *viz.* citral 'a' (geranial) 40-48 % and citral 'b' (neral) 30-37 %. Other chemical constituents *viz.* citronellal (1.1 %), methyl heptenone (1.9 %), geraniol (2.9 %), geranyl acetate (6.7 %), b-elemene (0.075 %) and b-caryophyllene (0.32%) should be minimum.

ii. **Citronella grass (*Cymbopogon winterianus*):** There are two types of citronella oil: Java type and the Ceylon type. The Java oil is much more valuable and useful as it contains about 80-85% of the total geraniol (including citronellal 37%, geraniol 25% and citronellol 18%). Whereas the Ceylon oil material for the extraction of several very important isolates (geraniol, citronellal) which can be converted into some of the most widely used aromatics (citronellol, hydroxyl-citronellal, synthetic menthol, esters f geraniol and citronellol etc.). The Java oil is higher priced and used in perfumery, toiletry (Masada, 1976). The Ceylon oil is lower priced and used in lower grades of soap, spray, distinfectants, polishes etc.

iii. **Palmarosa (*Cymbopogon martinii* var. motia):** Good quality oil should have high amount of free geraniol, lowest amount of geranyl acetate and presence of good number of esters (geranyl, neryl, prenyl etc.) in traces amount, which makes this oil unique (Pareek et al., 1981). The GC composition of good quality Palmarosa oil is geraniol (86-90%), geranyl acetate (5.5%), linalool (2.54%), terpenes and low boilers (1.29%) b-caryophyllene (1.02%), geranyl formulate (0.84%), nerol (0.50%) and higher esters of geraniol and high boilers (1.36%).

iv. **Mint/Pudina (*Mentha arvensis*)**: Total menthol content is the main constituent in *Mentha* oil. The oil is dementholized by freezing, rectification and other ancillary processes to separate part of natural/free menthol i.e. 1-menthol for use in flavour industry. The dementholized oil is used in flavour, pharmaceutical and confectonary preparations. Other chemical constituents of *Mentha* oil are: 1-menthone (6.8-22.2%) and menthyl acetate (14.1-24.4%). *M. spicata* oil contains 1-carvone (55-60%) which is used in flavour industry. *M. piperita* and *M. citrata* are used in lesser amount for flavour and perfumery industries, respectively. The essential oil obtained from the species *Mentha piperita* is known as peppermint oil. The oil is used for the flavouring of pharmaceutical and other preparations such as tooth paste, dental cream and mouth washes, gargles, cough drops, candy, chewing gum, confectionary, liqueurs and perfumes. In medicine, it is used as carminative and gastric stimulant.

v. **Basil/Tulsi (*Ocimum* species):** Major chemical constituents in the essential oils of some of the species available in India are *Ocimum basilicum* (Linalool 70-80%); *O. americanum* is of two types having major constituent of essential oil (i) Methyl chavicol 70-75% and (ii) Citral 75-80%. The oil of basil is used in flavouring of confectionary, food, condiments, dental cream, perfumery and mouth washes. It is also used in pharmaceuticals.

vi. **Khas/Vetiver (*Chrysopogon zizanioides*):** Vetiver oil contains a total of 150 sesquiterpenic compounds. It is evaluated for its total alcohol content, calculated as vetiverol, which ranges from 45 to 85 per cent in different cultivars and hybrids Guenther, 1977). Because of the high boiling points of its chief volatile constituents, and their high viscosity, the distillation of vetiver roots is somewhat difficult. Therefore, distillation of roots is carried out with live steam of 4 to 5 atmosphere pressure. The vetiver oil is extensively used in high quality perfumes and scents, cosmetic and soaps and also acts as a good fixative in perfumery formulation. Roots and root oil are used as carminative in flatulence, colic and as an anthelmintic and possess stimulant, diaphoretic and refrigerant properties.

3. Chemical components of essential oils and their therapeutic uses

The essential oils are a mixture of many compounds like acyclic, alicyclic, aromatic and heterocyclic components but some of them are simple and consist mainly one component. Some of the chemical components of essential oils and their therapeutic uses are given below:

i. **Terpenes:** A common terpene hydrocarbon is Limonene, which is found in 90% of citrus oils. Another terpene is pinene, which is found in pine oil. Terpenes may be slightly antiseptic, stimulating, and expectorant. Some may have antiviral or spasmolytic properties. Some diterpene may have a positive effect on the endocrine system.

ii. **Alcohols:** The alcohol group is formed when units consisting of one hydrogen atom and one oxygen atom attach themselves to carbon atoms. Alcohols tend to have good uplifting qualities; they are generally non-toxic with good antiseptic properties, and are often used in skin care products.

iii. **Aldehydes:** These are formed by the oxidation of alcohols and may often have a powerful aroma. Their chemical components may have a balancing, sedating, or antiseptic effect. The important aldehydes, found in lemon scented plants (lemongrass) are citral (geranial and neral), citronellal etc.

iv. **Phenols:** In essential oils, phenols are stronger than alcohols. They may often act as stimulants to the nervous and immune systems. Their overuse may cause skin irritation. Common phenols include eugenol, thymol and carvacrol- found in clove, thyme, or oregano plants, respectively.

v. **Ketones:** Only a few ketones are present in essential oils. Used in moderation, their effect is calming and sedating. They may help with balancing body fat, easing the flow of mucus and helping with skin rejuvenation and scar tissue formation e.g. menthone, thujone, and camphor.

vi. **Oxides:** The most important oxide is 1,8-cineol or eucalyptol. Eucalyptol has expectorant and mucolytic properties and is widely used in many pharmaceuticals.

vii. **Acids:** Acids and esters are a complex combinations of carbon, hydrogen, and oxygen atoms. Esters, which may have a fruity aroma, are effective for skin problems and may help in balancing the nervous system. Linalyl acetate is found in bergamot, lavender, or clary sage.

viii. **Lactones**: These molecules are too big to pass through a distillation process. Lactones are therefore usually present in oils prepared through expression methods or using concretes methods – like those processes used to produce jasmine oil.

4. Prospects

Keeping in view of chemistry and quality of essential oils, the incorporation of new aromas into perfumes and fragrances and the development of new products is critical to the success of the perfumer and flavour chemist. The difficulty for the agricultural researcher is to identify the types of aromas and chemical constituents desired by the perfumer and flavorist. The identification of new species rich in desirable essential oils may be promising route to pursue.

References

ASTA (1968). Official analytical methods of the American Spice Trade Association ASTA, Inc., Englewood Cliffs, NJ.

Guenther, E. (1972). The production of essential oils: methods of distillation, enfleurage, maceration and extraction with volatile solvents. In: Guenther, E. (ed.). The essential oils. History–origin in plants, production analysis. Vol. 1:85-188. Krieger Publ. Co., Malabar, FL.

Guenther, E. (1977). The Essential Oils. Vol. IV. Pub. By Robert E. Krieger Publishing Co., Inc., Huntington, New York. Pp. 20-65, 156-180.

Heath, H.B. (1981). Source book of flavours. AVI, Westport, CT.

Madan, V.K. (2008). In: Proceedings of National Seminar on "Recent Trends in Research on Spices and Aromatic Plants" held at CCS HAU, Hisar from Sept., 8-10, 2008, pp. 156-159.

Masada, Y. (1976). Analysis of Essential Oils by Gas Chromatography and Mass Spectrometry. Pub. by John Wiley & Sons, Inc., New York, pp. 13-22, 269-275.

Pareek, S.K., Maheshwari, M. L. and Gupta, R. (1981). Cultivation of Palmarosa oil grass. Indian Farming, 31(4): 22-25.

7

Use of Essential Oils in Aromatherapy

Introduction

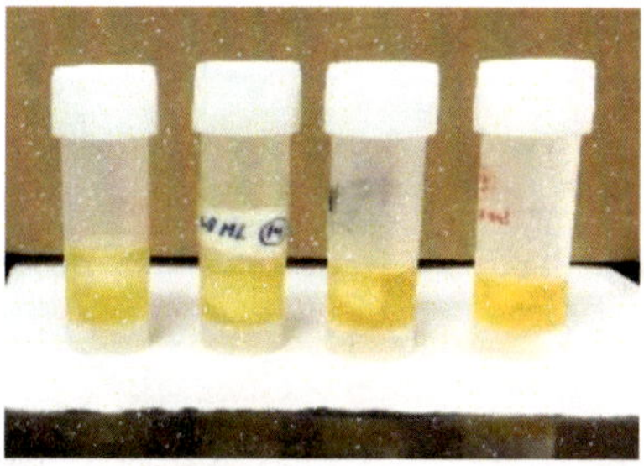

Essential oil obtained from aromatic plants is a mixture of several naturally arising biochemical compounds that occur in some plants which have scented flowers, leaves, bark, seeds, stem wood, or roots. As per the biochemical view, the active ingredients of essential oils are tranquil of oxygenated hydrocarbons and several other types of hydrocarbons which may be clustered conferring to their biochemical structure into esters, terpenes, aldehydes, alcohols, phenols and ketones etc. Carbon is considered as the basic element for all kinds of essential oils. Majority of essential oil compounds are made of C, H, O, N and S atoms. The atom of these different kind of elements are joined to each other to make the essential oil molecules. These biomolecules are made up of five carbon atoms is named isoprene. All essential oil compounds mainly consist terpenic compounds having two or more than two isoprene units (C_5). These Isoprene units (C_5) are generally attached to each other in head to tail fashion, hence 'isoprene rule' is followed. When two units of isoprene (10 C) are joint together, they form what is known as monoterpenes. Likewise, when three, four, five, six and eight units are united, they are called sesquiterpenes, diterpenes, sesterterpenes, triterpenes and tetraterpenes, respectively. These terpenes are colourless liquids and are generally lighter than water. They are highly refractive and insoluble in water. The 'terpene' term is usually restricted to hydrocarbon but on addition of oxygen in it, these are called terpenoids. These include menthol, cirtral, linalool, geraniol, carotenes, camphor, thymol, menthone etc. (Maheshwari, 1995).

Generally, the essential oil is extracted from aromatic plants through various methods but for production of essential oil for aromatherapy, the only suitable methods are those that are non-toxic. This contains cold pressing, distillation, and, in some cases, extraction by using the non-toxic solvents. An excellent quality essential oil for therapeutic use is formed through non-toxic procedures that reserve the life-giving vital essence of the aromatic herb. The essential oils also vary significantly in quality parameters as well as in therapeutic grade. There is confusion that vegetable oils are also essential oils. The vegetable oils evaporate when exposed to air; they are very fragrant; do not feel slickly or oily, and do not leave an oily residue. Vegetable oils, such as apricot oil, avocado oil, almond oil or olive oil, are not actually aromatic; they do not evaporate readily; feel oily and leave an oily remainder. Vegetable oils are also utilised in aromatherapy for diluting essential oil so that they can be applied safely to the skin.

Chemical components of essential oils and their therapeutic uses

The essential oils are a mixture of many compounds like acyclic, alicyclic, aromatic and heterocyclic component but some of them are simple and consist mainly one component. Some of the biochemical constituents of essential oil are listed below:

i. **Terpenes:** Generally, Limonene is a basic hydrocarbon, which is originated 90% from citrus oil. Another important terpene is pinene, it is mainly extracted from pine oil. The terpenes are considered expectorant, stimulating, and slightly antiseptic also. Some of these terpenes also have spasmolytic or antiviral activities. Some diterpenes may impart a progressive outcome on our endocrine system.

ii. **Alcohols:** The compounds of alcoholic group are made of one H atom and one O atom fasten themselves to C atoms. Alcoholic compounds have a tendency to have good uplifting abilities; they are considered as non-toxic compounds with decent anti-septic characteristics, and are regularly utilised in skin care items.

iii. **Aldehydes:** These compounds are made through the oxidation process of alcoholic compounds and generally contains a powerful aroma. These bio-components have a sedating, balancing, or anti-septic effects. Some important aldehydes, also found situated lemongrass, are citronellal and citral (geranial and neral), etc.

iv. **Phenols:** Among the essential oil components, phenolic compounds are robust than alcoholic compounds. They are found to be perform as stimulant to the immune and nervous systems. The overuse of these compounds may leads to skin irritation. Common phenolic compounds are carvacrol, eugenol, and thymol – found in oregano, clove, or thyme plants, respectively.

v. **Ketones:** In the essential oil, the presence of ketones is found very less. The ketones are used in self-control, their effect is sedating and calming. They also help in easing the flow of mucus, balancing body fat, and also helps in scar tissue development and skin rejuvenation. Examples of ketones are camphor, methone and thujone.

vi. **Oxides:** The most important oxide is eucalyptol or 1,8-cineol. Eucalyptol has mucolytic and expectorant characteristics and is generally utilised in pharmaceuticals industries.

vii. **Acids:** Acids and esters are the complex structures of C, H & O atoms. Esters possess a fruit like aroma, and found operative for skin related problems and also found beneficial in balancing the nervous system. Some plants such as clary sage, lavender, bergamot, are found to have Linalyl acetate, an acidic compound having medicinal value.

viii. **Lactones:** These biomolecules are very large to go through a distillation process. Therefore, lactones are usually found avaiable in the oils prepared by the expression procedures or using concretes procedures – like those procedures utilised to harvest jasmine oil.

The **Gas Chromatography–Mass Spectrometry (GC-MS)** analysis is a suitable technique to analyse and identify individual constituents of essential oils.

Aromatherapy

The essential oil obtained from aromatic plants used for inhalation and external application on the body to enhance or restore the health, wellness and beauty.

The elementary objective of Aromatherapy is to combine the systematic research realizations of man with his untaught understanding for the illness management with the supreme active and valuable natural essential oils. According to the law of nature, the basic principle of Aromatherapy is to make stronger the self-healing procedure by preventative method and indirectly stimulate the immune system of our body. The area of biological activity of aromatherapy is fairly widespread, ranging from penetrating deeply in the tissues for therapeutic activities to the extreme subtlety of distinctive fragrance. The use of essential oils is very simple to increase heath as well as the quality of life.

How do essential oils work?

Essential oils function in synchronization with the body to balance and normalize. They impart definite effects that can be measured, and required to be adapted according to the needs of different persons. Utilised for their antagonistic effects on microbes and antiseptic properties. The naturally obtained essential oils are found less toxic when compared with artificially synthesized anti-biotics but also seems to sustenance life (eubiotic) by functioning with the human beings own natural healing abilities (by which means true healing takes place). Some essential oil compounds found in Roman chamomile, have antiseptic, wound healing, anti-inflammatory and cell regeneration properties as well as anti-fungal effects making them the final active ideologies for universal skincare purposes.

What are the effects?

Essential oils may cause direct or indirect effect on the human physiological systems. For illustration, two drops of peppermint given orally found able to improve digestion; and inhalations of mucolytic oils found effective to relieve the respiratory problems. It also has antiseptic and soothing effects, and found effective to treat minor skin problems. The application of some of the essential oils on body skin found able to produce vaso-dilation and imparts warming effect on underlying muscles, but this effect of the oil was found on superficial tissues only, and it is not entering into the systemic circulation through the skin. Moreover, it also helps in the relaxation on the brain and the consequent sedating or stimulating of the nervous system, essential oil seems to be indirectly increase and reduce the blood pressure and perhaps helps in regularization of hormonal secretions. Due to direct connection of olfactory glands to the brain, conveying electrical impulse directly into the limbic system, it also imparts positive effects on mental states and emotions. Perception of perfumes also have a major impact on thinking, learning, memory, emotions, and feeling of mankind. As therapeutic representatives, essential oils function likewise to tranquilizers but in a delicate organic way. The most of the aromas help to calm the nervous system and uplift spirits. For instance, basil, rosemary and peppermint are working as uplifting and stimulating agents; lavender calming and sedative; and ylang-ylang and jasmine are euphoric or exciting.

How are essential oils used?

Essential oil direct inhalation imparts psychological effects via olfactory nerves to the limbic system and gives simulative or sedative effects on body. Furthermore, physiological effects were noticed due to the fastest way into the bloodstream. Inhalation in of essential oil found to have the greatest benefits regarding to respiratory problems and it can be utilized through sniffing drops on a tissue or by inhaling near a diffuser with glass nebulizer. Local uses of diluted essential oils (2-10% blended with vegetable oil) on various points (chakras, spinal nerves, meridians) is found useful for certain problems. Moreover, full-body massage is also imparting significant effect, providing relaxation as well as a physiological action through the nervous system. Even though, it is also reported by researchers that many physiological systemic benefits are as a result of essential oils penetration in the blood through application on skin, little confirmation backs this up as a major route of absorption. In addition, many of the therapeutic claims are "borrowed" from herbalism, or are really the effect of ingesting the herbal extract. Knowledge and common sense regarding to use of the essential oil as a particular medicinal is required. Moreover, knowledge of safety features is compulsory before using any kind of essential oil on the body skin, as literature tells us that some of them are found sensitizing or irritating and even toxic also.

Patch test: Some persons are sensitive to about essential oils; therefore, it is required to carry out patch test on the user to know about the reaction. Take 1-2 drops of essential oil or mixture of essential oils which you want to use and smear behind the ear or on the wrist and observe any reaction at least for 30 minutes. If the skin demonstrations any itching or irrigation, redness, you should escape the usage of this particular essential oil.

Storage of essential oils: The majority of most essential oils have shelf-life around 12 months, if stored in proper way. The essential oils needed to be stored at cool places, where sunlight does not penetrate. The bottles having essential oils, only to be opened as and when it is required and just after use tightly close the cap, because its exposure to air may fasten the speed of deterioration of the oils, including essentials oils.

Aromatherapy oils and their uses

- **Eucalyptus:** To treat the respiratory ailments by enhancing deep breathing, Eucalyptus oil is used by aroma therapists. This oil is suggested for muscle aches, sprains, or pains.
- **Geranium:** It is useful to treat the nervous tension, and PMS symptoms, stress-related conditions. It also repairs cracked capillaries; fights to correct eczema, cuts and burn.
- **Jasmine:** It helps in warming, intoxicating, relaxing, and great aphrodisiac. Jasmine oil is in use from long time for females at the time of baby birth as a supportive and relaxing agent. It is most valuable product for the irritated skin, sensitive and dry skin.
- **Lavender:** It utilized in the treatment of sprains, rheumatism, abdominal cramps, respiratory problems, sun-damaged skin insomnia, depression, burns, tension-related complications, and various kinds of skin contaminations.
- **Patchouli:** Its oil imparts a valuable effect on the body skin, against the insect bites and infections. It is also applied against the addictions and stress related complications.
- **Peppermint:** It is utilized for the treatment of nausea, asthma, craps, colic, colds, fainting, fevers, mental fatigue, headaches, and flatulence.

Its oil is also found helpful to eliminate apathy, migraines, shock, nervous stress, faintness and vertigo.

- **Rose:** Essential oil of rose is used in stress related conditions, headache, insomnia, nervous tension, and nausea. Also found useful in case of broken capillaries, dry skin and poor circulation.
- **Rosemary**: Oil can be effective for circulation problems for mental fatigue, also acts as a muscular pain reliever and can help to treat respiratory and skin problems.
- **Sandalwood:** It is great aphrodisiac relaxing and warming nature. Sandalwood is cherished in skin-care for its skin healing and moisturizing assets.
- **Sweet Orange**: Its oil is used by Aroma therapists to treat poor circulation, obesity, muscular tension, colds, bronchitis, water retention, constipation, and stress-related symptoms.
- **Tea Tree:** It has strong anti-fungal, anti-bacterial, and disinfectant properties.

Avoid the following oils in case of pregnant woman

- Basil, Cedar wood, Clary Sage, Jasmine, Fennel, Juniper, Lavender, Rose, Marjoram, Myrrh, Sage, Rosemary and Thyme.

Avoid the following oils in case of Epilepsy patient

- Camphor, Fennel, Rosemary, Sage and Hyssop.

As we have discussed that in the aromatherapy, only the utmost superiority of essential oils would be used. The most of essential oils formed in the markets are utilised in the food flavour and fragrance industry. Essential oils available in the different shops, are mixed with comparable resemblance in essential oil composition, chemical and synthetics, as well as extenders such as dipropylene glycol. Therefore, it is wise to procure the essential oils only from a reputable aromatherapy source.

References

Guenther, E. (1977). The Essential Oils. Vol. IV. Pub. By Robert E. Krieger Publishing Co., Inc., Huntington, New York, pp. 20-65, 156-180.

Madan, V. K. 2005. In Course Compendium of Winter School on “Advances in Medicinal, Aromatic and Under-utilized Plants Research” held at CCS HAU, Hisar from Sept. 29 to Oct., 19, 2005, pp. 240-245.

Maheshwari, M.L. (1995). Quality and Chemistry of Essential Oils. In: Advances in Horticulture. Vol. 11- Medicinal and Aromatic Plants Eds. K.L. Chadha and Rajendra Gupta. Published by Malhotra Publishing House. New Delhi. Pp. 145-197.

Masada, Y. (1976) Analysis of Essential Oils by Gas Chromatography and Mass Spectrometry. Pub. By John Wiley and Sons. Inc., New York, pp. 13-22, 269-275.

Pareek, S.K., Maheshwari, M.L. and Gupta, R. (1981). Cultivation of Palmarosa oil grass. Indian Farming, 31(4): 22-25.

8

Post Harvest Management in Aromatic Plants

Aromatic plants are one of the major natural resources of our country occurring in diverse eco-system. Lately, many people due to rare chances of side effects prefer herbal medicines/remedies and their safer products. Aromatic plants sector of our country is not well organized which needs to be given special attention in States as well as National level. At present in India, total trade of Ayurveda and other herbal products is estimated to be around 5000 crores, which indicates that medicinal and aromatic plants have good potential. Presently, 90 percent of traded medicinal plants are generally collected from wild and rest from cultivation. These are either used fresh or in the form of crude drug, either dried whole plant or their part/parts such as root, stem, wood, bark, leaves, flowers, fruits, seeds and their products. The major producers of medicinal products in the world are China, Brazil, Bulgaria, Poland and India.

Though India has rich medicinal and aromatic plant diversity, yet China is the largest exporter of medicinal plants earning huge amount (US $ 5 million) from herbal trade due to large scale cultivation, standardization of quality control, safe and regular suppliers to the international market. Despite being home for several Traditional System of Medicines and aroma industries, the export of herbal based products from India is below US $ 100 million. According to WHO estimates, the herbal and aroma market is growing at the rate of 7 percent annually. The major export market for India is USA, Japan, Germany, France, Australia etc.

There are several constrains in this sector, like, unscientific and unsustainable collection from natural habitat, habitat destruction, illegal collection, low return, no uniformity in rates, insecurity of growers, lack of structured marketing and processing unit in the adjoining areas and regions. Even though, aromatic plants sector would be successful, if backed by farmers support, for which they need to be assured, safer and higher return. Simultaneously, training on conservation, collection, identification, cultivation, harvesting, drying, grading, storage etc. should be given through technical persons. Therefore, linkage among farmers, traders and processing units should be facilitated to assure continuous supply of raw material through buy back agreement with predetermined price or fixed by the government.

The post-harvest management and processing of aromatic plants consists of asset of procedures or operations to which the plant material is subjected prior to its marketing. The main purpose behind these processes is to stabilize the material for transport and storage, ensure its free from foreign organic matter, substitutes and unintentional adulterants and proper care of its pharmaceutical elegance to facilitate its presentation in a marketable form. The post-harvest process must be designed scientifically and carried out in such a manner that the chemical and medicinal contents of the material are not affected qualitatively as well as quantitatively. A proper processing is not only important for obtaining a quality herbal raw material but also for value addition in it.

Quality management system should be followed in the total operations, starting from cultivation to the quality control and packaging across various intermediate processes like cultivation, storage, extraction, formulation, processing and standardization of aromatic plant in the field. Proper documentation at each stage of preparation of different formulations like powder, extract, liniment, liquid, tablets, capsules has global importance. The analytical methods used in quality assurance (QA) need to be refined, based on special requirements and guidelines of the legislation in different processes of herbal drug manufacturing quality, purity and conformity with uniformity and consistency.

Therefore, post-harvest management, processing and marketing are of great significance in aromatic plants, as several types of biological activities are under process even after harvesting. Due to lack of post-harvest knowledge and safe storage practices, a large quantity of product gets spoiled in their major chemical constituents due to improper drying of the end produce, and as a result sold at much lower price in the market. Therefore, proper drying of the material harvested to reduce the moisture up to minimum desired level before storage is of great significance.

9

Celery/Ajmod (*Apium graveolens* L.)

Introduction

It is a common vegetable crop of *rabi* season which belongs to the family Apiaceae (Umbelliferae). Due to prevalence of protandry condition, the crop is cross pollinated in nature and self-pollination within individual flower is restricted. It comprised of three distinct cultivated varieties grown over the world for various economic purposes and classified as below depending upon usefulness of different parts.

Common Name	Species	Characters and usefulness
Celery/ Ajmod	*Apium graveolens var. deluce*	It is very succulent plant with solid petioles utilized as salad.
Root celery/ Celeriac / knob celery and turnip rooted celery	*Apium graveolons var. rapaceum*	Well adapted to marshy lands. Mainly cultivated in Europe and some parts of Asia. Developed hypocotyls/root tissue are eaten in cooked form in soups and nettled on salad.
leaf celery / Smallage	*Apium graveolens var. secalinum*	Its Leaves have hollow and slender petioles are consumed as salad. Extracted essential oil is used in pharmaceutical and food flavour industries. Its leaves are also used for medicinal, condiment, and garnishing purposes.

Origin

It found is globally distributed in New Zealand, Australia, South Africa, Mediterranean basin, and South America, whereas Argentina and Chile are the richest sources. Habitat is extensively ranging from Egypt, Algeria, and Abyssinia to Sweden.

Other Uses

1. Used as vegetable and as mixed salads or an appetizer.
2. Often stuffed with cream cheese or peanut butter.
3. Petioles are cut into pieces and served as cooked vegetable.
4. Mostly used in vegetable juices, soups and stews.
5. Volatile oil is used as natural preservatives triggered research on anti-bacterial, anti-fungal and anti-oxidant.

Genotypic variability and Cytogenetics

About 20 species under genus *Apium* are reported, which are annual as well as perennial. *Apium graveolens* species is diploid with 2n = 22 chromosomes, and basic set of chromosome i.e. genome as x = 11. This is the only commercially viable cultivated species in India. The other prevalent species in the world are:

***Apium* Species**	**Status**
Apium graveolens	Commercially cultivated
Apium nodiflorum	Wild in nature and observed in damp/marshy lands in European, Egyptian and Asian countries.
Apium chilense	-do-
Apium panul	-do-

Floral biology

It is an annual glabrous herb with erect stem, pinnate compound leaves with long stalks. Inflorescence bears greenish white flowers in compound umbels. Flowers are having five petals in corolla and five stamens in androecium. Its seeds are developed from two carpels of compressed structure; and have very small dark brown colour with agreeable odour and pungent taste.

Chemical compositions of crop

All plant parts *viz.* sleave, stalk and seeds; and essential oil obtained from seeds are important from industrial point of view. The quality compositions of leaves and stalk differs at harvesting of seed setting stage. However, chemical composition of leaves is better over the stalk as given below:

Composition	**Stalk (%)**	**Leaves (%)**
Moisture	93.3	81.3
Carbohydrates	3.5	8.6
Fat	0.1	0.6
Proteins	0.8	6.0
Calcium (Ca)	0.03	0.23
Phosphorous (P)	0.04	0.14

Iron (Fe)	4.8 mg/100 g	6.3 mg/100 g
Vit. A	Nil	800-7500 IU
Vit. C	6 mg/100 g	62 mg/100 g

Chemical composition of seed oil

Particulars	Valve	ISI Specification
Physico-chemical parameters:		
Colour	Pale yellow	Pale yellow to light brown & pale-green
Specific gravity	0.850-0.895	0.871-0.910 (27°C)
Refractive index	1.478-1.486	1.4765-1.4865
Optical rotation	+ 65-82 degree	+ 50° to 80°
Acid valve	15-40	Max. 3.5
Odour	Spicy characteristic of celery seed	Persistent spicy & typical of celery seed
Chemical constituents:		
d-Limonene	60 %	-
d-Selenene	10 %	-
*Sedanoic acid anhydride	0.5 %	-
*Sedanolide	2.5-3.0 %	-

*Chemical responsible for aroma.

Breeding methodology

i. **Hybridization:** Tiny flowers are organized in inflorescences called as umbels. Though highly prone to self-pollination, yet cross-pollinated due to protandry conditions. Hybrid breeding is the major objective with increased yield vigour, high uniformity and disease resistance. Therefore, open-pollination resulted in wide variation, followed by further selections over years. Lack of emasculation possess difficulty in obtaining F_1 hybrid seeds. Hence hybrid cultivars availability is less in the market, as compared with open pollinated varieties.

ii. **Male sterility**: It is either determined by genetic (nuclear gene) or cytoplasmic (cytoplasmic organelles) in nature. Genetic male sterility (GMS) segregates and can be propagated vegetatively. Cytoplasmic male sterility (CMS) and cytoplasmic genic male sterility (CGMS) are determined by mitochondrial encoded genes. Therefore, these are inherited maternally and economically advantageous. Quiros *et al.* (1986) reported first male sterile line of Iranian accession (P 1229526), governed by a single recessive gene (*ms-1*).

CMS was first time exploited in celery line '021542' serving as a female parent and chervil '022047' as a male parent subjected to manual control hybridization. Male sterile plants were selected from the F_1 group as a cytoplasmic male sterile source and was subjected to fifth backcross generation to obtain a cytoplasmic male sterile line '0863A'; and a backcross male parent corresponding cytoplasmic male fertile line as a maintainer line '0863B'. The sterile line '0863A' and a selected male fertile parent line were subjected to field planting in a cross-breeding farm with an excellent isolation condition in a ratio of 3:1. After pollination, male parent plants were uprooted completed, and seeds picked from sterile line plants after maturity of the seeds as hybrids. The cytoplasmic male sterile line bred by the method had stable heredity, high combining ability, simple cross-breeding process, low cost and high quality hybrids, used for the large scale cross breeding production.

Dawson (1993) reported that CMS was introgressed from a wild unidentified unstable celery source plant resulting in some of hybrid celery and celeriac cultivars.

Gao *et al.* (2006) find out a new male sterile plant **(01-3A)** under GMS while visiting the commercial celery crop of the inbred line **(01-3)**. This was used to create hybrid **'Jinqi 2'** and subsequently reported that the male sterility was controlled by two nuclear recessive genes.

Bruznican *et al.* (2017) reported that CMS system could be transferred by asymmetric protoplast fusion as an efficient protocol for protoplast regeneration.

iii. Breeding for Disease Resistance:

a) **Late blight (*Septoria apiicola*):** A highly detrimental leaf disease under poorly managed conditions. Resistance against this was introgressed from another crop known as parsley (*Petroselinum hortense*). Subsequently, resistance was identified in wild species *viz. Apium chilense*, *Apium nodiflorum* and *Apium panul*. But the two latter species, when crossed to celery, resulted in somewhat less resistant F_1 hybrids than their resistant parentages, demonstrating inadequate dominance.

b) **Wilt (*Fusarium oxysporum*):** Resistant celeriac landrace when crossed with the susceptible celery cultivar resulted in resistant line **UC-1**, which was later on backcrossed to elite varieties. As a result, three resistant germplasm lines *viz.*, **UC26-1, UC10-1 and UC8-1** were released as commercial cultivars.

Selection through soma-clonal variation through regeneration from suspensions of cells resulted in some resistant lines. Later on, after five generations of self pollination and selection a stable resistant line was obtained.

c) **Early Blight (*Cercospora apii*):** It required an intensive as well as costly disease management is required to control this highly transmissible disease. Subsequently a strain **'Floribelle M9'** was found with superior resistance compared to other resistant cultivars. Further, resistance was traced to a Turkish accession of celeriac cultivar and **FBL 5-2M** was developed.

d) **Mosaic Virus:** Celery mosaic virus (CeMV) transmitted through aphid, and it was found to be governed by a single recessive gene (*cmv*). Further, markers linked to this gene were identified and facilitated introgression for resistance in other celery varieties.

iv. **Breeding for Insect Pest Resistance:**

a) **Beet armyworm (*Spodoptera exigua*)**: Resistance against this pest was identified in another species *Apium prostratum,* which was otherwise difficult to control by chemicals.

b) **Leaf miner (*Liriomyza trifolii*):** High yield loss and decreased marketability of the produce is caused by this pest. Two wild species *viz. A. nodiflorum* and *A. prostratum* were reported to have natural resistance against this pest. Subsequently, celery was crossed with *A. prostratum* in a series of backcrossing, and found that F_1 hybrid and some of the backcrossed lines maintained the resistance as observed in the wild parental accession.

v. **Breeding for Nematode resistance:** *Meloidogyne incognita* and *M. javanica* were found to cause root-knot disease in tropical areas, whereas *Meloidogyne hapla* in temperate regions. Later on, out of five celery cultivated varieties tested for resistance to *M. hapla*, and two lines **'Green Boy'** and **'Dutchess'** were developed with considerably lower numbers of adult nematodes.

vi. **Breeding for Late Bolting:** Bolting mainly starts during the first year of plant development, as premature bolting happens due to stress induced suppression during vegetative growth, resulting in chief impression on the crop yield as well as on its quality as the inflorescence stalk (bolting) consumes the energy produced in the leaves and petioles. An extended cold period dependence is genetically determined and highly variable.

Generally, biennial varieties displayed low to high bolting resistance in comparison to annual celery cultivars. Therefore, European celery and celeriac varieites are biennial having sufficient resistance even in the first growing season.

vii. **Breeding for Abiotic Stress and Phyto-remediation:**

a) **Salt tolerance:** Increased mannitol biosynthesis controlled by mannose-6-phosphate reductase **(M6PR)** resulted into salt tolerance. The results further suggested that ***AgVIP*1** may be included in the celery plant stress response mechanism and could work as a marker for reporting the unknown resistance position of some celery varieties, as expression pattern was found to be closely associated with the genetic background of the cultivated varieties.

b) **Heat stress:** The expression of the ***HEAT SHOCK FACTOR B2*** **(*AgHSFB2*)** gene conferred security in cultivars **'Liuhe Huangxinqin'**, **'Jinnan Shiqin'**, and **'Ventura'**. This gene differed regulation among the cultivated varieties with **'Ventura'**, accruing sooner and greater transcript levels response to heat.

c) **Cold stress:** ***DEHYDRATION-RESPONSIVE ELEMENT BINDING1*** and ***2*** **(*AgDREB1*** and ***AgDREB2*)** transcription factors upgraded resistance to cold-stress in the transgenic plant of *Arabidopsis*. Therefore, have potential for varietal screenings in breeding programs against cold.

d) **Heavy metal resistance:**

- **Cadmium, manganese and zinc:** The different cultivars of celery and celeriac have expressed their potential regarding to accumulate these heavy metals such as Cd, Mn and Zn in petioles and leaves grown on soils contaminated with heavy metals within the permissible levels. It was found that out of twenty-seven (27) *Apium* accessions from China, only one line gathered the lowermost amounts of lead and cadmium in the contaminated field.

- **Boron:** Regarding the uptake of boron, the cultivars **'Emerson Pascal'** (an efficient boron user) and **'S48-54-1'** (susceptible to low boron) were examined in low and high concentrations of boron. The results revealed no difference in boron uptake between the two cultivars. But cultivar **'Emerson Pascal'** accumulated boron in shoots; and **'S48-54-1'** in roots of plants. Thus selection of cultivars based on their capacity to accumulate boron in different organs avoiding toxicity limit in the food chain can be done.

- **Arsenic, cadmium and lead:** Concentration visible to environmental pollution revealed that the aerial plant parts can accumulate these heavy metals. Cadmium and Arsenic can be sketched back in the hypocotyl of celeriac plants, unlike lead. Therefore, *Apium* could be cultivated as a bio-marker for environmental pollution, as it accumulated more pollutants than other crops. Hence improvement could helpful in development or identification the new cultivars that absorb the least pollutants (heavy metals) in the edible parts of the crop plants.

Suggested questions

1. What is scientific name of Celery/Ajmod.
2. Write 2-3 interspecific/genotypic variability.
3. Write the family to which Celery crop belongs.
4. Write the names of three cultivated varities of Celery.
5. What is the nature of pollinated and why?
6. What are the major chemical constituents of celery leaves?
7. What is the ploidy level and somatic chromosome number?
8. Write different uses of celery leaves.

10

Chamomile (*Matricaria chamomilla* L.)

Introduction

Chamomile is generally known as German or Hungarian chamomile. Its herb is commonly used for both medicinal as well as aromatic purposes. It was introduced in India during late fifties by Regional Research Laboratory, Jammu (Handa *et. al.*, 1957) and subsequently in different parts of India. It was successfully grown over saline soils at CIMAP, Lucknow and later at several other locations. It was also introduced at YSPUH&F at Solan (H.P.) under AICRP on M&APs and several parameters were identified for better utilization of inputs for higher production of herb. Its flowers also contain an essential oil with multiple uses.

Origin

It is native to Europe and mainly cultivated in the regions of northern, central and eastern European countries, Mediterranean, N. Africa, United States, Germany, France, Hungary and Russia. It is found grown in wild state in eastern and southern Europe, West Asia and North Africa.

Uses

- Flower heads are primarily utilised for nutraceutical and medical purposes.
- Widely used to treat different kinds of diseases with anti-oxidant, anti-bacterial, anti-fungal, insecticidal, anti-parasitic, anti-inflammatory, anti-diabetic, and anti-cancer effects.
- Facilitating its application in food preservation, phyto-sanitary control, surfactant and anti-corrosive agent.
- Used as sedative, anti-spasmodic, anti-septic, and anti-emetic, curing infections, neuro-psychiatric, respiratory, gastro-intestinal, and liver disorders.

- Used in tea preparation
- Tinctures preparation from dried flower heads with ethanol and water in 1:5 ratios.

Floral Biology

It is an annual herbaceous plant having plant height of 60-90 cm. The central stems are more or less erect, with semi-erect side branches. Chamomile flowers are cross pollinated owning to protandrous condition i.e. anthers mature 24-48 hours earlier than stigma receptivity. Flower head is solitary and borne on a semi-spherical of conical hollow receptacle surrounded by 2-3 rows of small imbricate bracts, producing 40-50 seeds with yellow to light brownish in colour. It belongs to family Asteraceae.

Genotypic variation

Chamomile is ascribed to three genera as given below:

Genus/Species	**Common name**
Matricaria chamomilla L.	Hungarian or German chamomile
Anthemis nobilis L.	Roman chamomile
Ormenis multicaulis Braun-Blanquet and Maire	Moroccon chamomile

Cytogenetics

Matricaria chamomilla L. (syn. *Chamomilla recutita* L.) is found in diploid state having 2n = 2x = 18 chromosomes.

Besides, tetraploid chamomiles with 2n = 4x = 36 chromosomes, obtained after colchicine treatment. Tetraploids differed from diploids for several biochemical and morphological traits e.g. plant height, seed weight, essential oil yield, inflorescence size, dry inflorescence weight, bisaboloids and chamazulene content. Therefore, farming of tetraploids was preferred over diploids.

Genetic variability

Sufficient variability for essential oil content as well as its bioactive compounds as above were reported by various workers in different commercial varieties of chamomile populations.

Taviani *et al.* (2002) reported that genetic diversity among wild chamomile populations. Gosztola *et al.* (2006) evaluated wild populations collected from Hungary for its natural habitats; and suitable ones to produce high quality herbal medicines; and selection of genotypes for the development of new cultivars.

Franke and Schilcher (2007) reported genetic variability for various morphological traits i.e. big inflorescence, uniform plant height and high production; qualitative traits e.g. high chamazulene and *α*-bisabolol content; and resistance to biotic and abiotic conditions.

Okon *et al*. (2013) exploited feasibility to obtain chamomile plants with good quality essential oil through the exposure of new genetic resources of diversity from wild native populations; and making their required characters into the new commercial cultivars.

Otto *et al*. (2015) pointed out the necessity of new plant genetic resources of chamomile with superior characteristics; and their use for achieving different breeding goals.

Otto *et al*. (2017) further reported that most of the high-performed tetraploid chamomile cultivated varieties are genetically similar, suggesting further broadening of chamomile's diversity by including diverse germplasm lines in the upcoming crop improvement programme.

Correlation and Path analysis

Oil content had a positive correlation with dry flower yield; and also had large direct and indirect effects of number of flower heads, days to flowering and flower size. On the other hand, oil content was negatively correlated with days to maturity, indicating the importance of early and optimal flower harvesting stage for higher oil yield.

Breeding Methodology

Small flies are the major pollinating agents for cross pollination. Higher dry flower yield, synchronous flowering, high oil content and better quality blue essential oil are the major objectives. Although more than 80 percent seed formation has been reported under both selfed and open pollinated conditions. But studies on the development of advance generations and inheritance of different trait are limited. Studies on characters responsible for chemical components contributing towards high quality of essential oil revealed that a single recessive gene is responsible for control of chamazulene and (−) -α-bisabolol content.

i. **Honeycomb methodology:** This method followed a systematic record arrangement to handle the soil heterogeneity and stress minimization. Honeycomb design compared diverse genotypes by allocating each genotypes systematically and uniformly according to a hexagonal organisation in the entire field.

ii. **Germplasm Conservation:** Collection and conservation of germplasm at seed banks and field gene banks are considered easy method to avoid excess collection from wild sources and loss of genetic resources.

Polyploidy

Development of polyploidy breeding and selection from local landraces are important for evolution of variety and genetic improvement for higher flower and oil production. Tetraploids developed through colchicine treatment of seeds resulted in higher flower diameter, flower weight, plant height and increased flower-pick length. Though diploids and tetraploids were found to be statistically *at par* for oil content.

Triploid developed for commercial cultivation using diploid and tetraploid parental crossing depended on the availability of male sterility and self-incompatibility.

Biotechnology

Biotechnological approaches are important for fingerprinting of genotypes and to explore relationships of species.

Terpenoid analyses through biosynthesis pathways worked out in both *M. chamomilla* and *Chamaemelum nobile* by gene co-expression networks discovered that higher terpene synthase expressions correlated positively with sesqui-terpenoids in German chamomile. Therefore, higher volatile compounds of *M. chamomilla* (German chamomile) increased resistance to pathogens due to ten times more unigenes responsible plant-pathogen interactions in comparison to *Chamaemelum nobile* (Roman chamomile).

Hairy roots transformation system revealed the over appearance of sesquiterpene synthase genes found responsible for the accumulation of γ-muurolene in hairy roots.

Varietal development

The only variety **Soraksar-60** (Ex-Hungary) attained promise and cultivated with success under AICRP on M&APs at YSPUH&F, Solan (H.P.) centre.

Chemical composition of essential oil

About 120 constituents in the essential oil of chamomile are generally terpenoids, *viz.* α-bisabolol and its oxides-A and B, chamazulene, β-farnesene and bisabolone. Moreover, the extracts are ruled by phenolic components, containing flavonoids, coumarins and phenolic acids.

Production Technology

Climate: It is widely adaptable and grown in both *rabi* (tropical & sub-tropical zones) and *kharif* season (temperate regions). Slightly cold temperature during early growth stages and warm dry weather at the time of maturity is the most suitable for higher production. At flowering and before maturity temperature above 30°C reduce the oil content in the produce.

Soil: Can be sown over a wide range of soil types but slightly heavy soils with high humus and pH ranging from 8.0 to 9.5 are preferred.

Nursery raising: Well prepared nursery beds are made for raising seedlings during October/November and transplanted in well pulverised and levelled fields during November/December having seedlings height of 5-8 cm at 30 x 30 cm spacing.

Fertilizer requirements: Well rotten farm yard manure (FYM) with 20-25 t/ha is required to be applied at least 2-3 weeks before transplanting of seedlings. Nitrogen (N) is split in two doses at the rate of 20 kg/ha (transplanting) and 40 kg/ha after 30 days, while phosphorous (P) and potash (K) at the rate of 30 kg/ha at the time of field preparation (Johri *et al.*, 1991 and 1992 a, b).

Inter-culture operations: 3-4 irrigations are supplied depending on field soil type and prevailing weather conditions. Applying irrigation at critical stage of flower development resulted in higher oil content and oil yield.

Harvesting and Yield: Time of harvesting is very critical for oil yield and its quality. In plains, flowering starts from February to middle of the April. The full bloom stage of flowers is the best for picking at 10-15 days' interval, depending upon the duration of flowering. Average fresh flower yield of 60 q/ha can be obtained from optimum field conditions. In view of high moisture content in freshly harvested flowers (60-85%), their rapid shade drying is very much required to have an average dried flowers yield of 10-15 q/ha.

Oil yield: Essential oil having blue colour with strong smell can be obtained from air dried flowers after steam distillation, which mainly consists of high boiling constituents, including paraffin. The oil content in dried flowers may vary from 0.3-1.3 percent depending upon the environment, including strain, location, soil fertility and temperature etc. Under sandy-loam soils, the oil content may vary from 0.54-0.84 percent, considered as quite good recovery. Further, thc oil quality was found to be good (high bisabolol and other compounds), when flowers are picked up at full blooming stage.

Post-harvest management

Essential oil of better quality can be obtained from fresh flower heads in comparison to dried flower heads owing to reduced chamazulene and bisabolol contents. Drying at lower temperature and shade-drying resulted in higher oil content than dried under higher temperatures in oven, microwave and sun drying.

References

Johari, A.K., L.J. Srivastava, J.M. Singh and R.C. Rana. 1991. Indian Perfum., 35 (2): 93-96.

Johari, A.K., L.J. Srivastava, J.M. Singh and R.C. Rana. 1992 a. Indian J. Agron., 37 (2): 302.304.

Srivastava, L.J. and A.K. Johri. 1995. In: Advances in Horticulture, Vol. 11, Edited by K.L. Chadha and Rajendra Gupta, Malhotra Publishing House, New Delhi, pp. 795-804.

Suggested questions

1. What is scientific name of Chamomile
2. Write three genera to which chamomile is ascribed.
3. Write the family to which chamomile crop belongs.
4. What is the released variety of chamomile?
5. What is the nature of pollinated and why?
6. What are the major chemical constituents of essential oil?
7. What is the ploidy level and somatic chromosome number?
8. Write different uses of chamomile essential oil.

11

Citronella grass (*Cymbopogon winterianus*)

Introduction

It was first introduced from Java by National Botanical Research Institute (NBRI), Lucknow in 1959. Later on Regional Research Laboratory (RRL) of CIMAP, Bangalore started research work from 1961. Presently, it is cultivated in Karnataka, Assam, Maharashtra, TN, WB, AP, Southern Gujarat and UP. Citronella oil is produced from two species, *viz.*, *Cymbopogon winterianus* (Java type) and *C. nardus* (Ceylon type). About 300-350 tonnes of oil is produced annually in India, as against 7000 tonnes in the world.

Origin

It is reported to be originated from Borneo, Java and Sumatra

Genotypic variability

The following two species are generally available in India.

Name	Area of growing	Oil colour	Oil odour	Quality parameters
Java citronella (*Cymbopogon winterianus*)	Indonesia, India, Ceylon, Sumatra	Clear, Mobile, Light yellow-Brownish	Aldehyde	Total Geraniol: 80-85% Citronellal: 37% Geraniol: 25% Citronellol: 18%
Ceylon citronella (*Cymbopogon nardus*)	Sri Lanka	Pale yellow-yellowish brown	Citronellal	Total geraniol: 55-65% Citronellal: 7-15%

Uses

Despite production of synthetic isolates in large scale, there is still a widespread preference for natural citronella oil because of its unique olfactory and stable technical properties, which are vital in blending perfumes and compounding industrially important essences. The oil is mainly used in perfumery, soaps, detergents, industrial polishes, cleaning compounds, mosquito repellent products and other industrial products.

Cytogenetics

Species	Chromosome No. (2n)	Reference
Cymbopogon winterianus (Java type)	-	-
Cymbopogon nardus (Ceylon type)	40	Gould and Satertorm, 1974

Genetic variability

Yadav *et al.,* 2012 studied genetic variability among 6 genotypes for oil yield and oil content. They reported that fresh herb yield per plant varied from 73.3-266.7 g with a mean of 183.1 g and HJ-1 recorded the highest fresh herb yield per plant (266.7 g). Oil content on fresh weight and dry weight basis were recoded to vary from 0.40-0.82 % and 1.29-2.41 %, respectively; and HJ-3, HJ-2 and HJ-1 were having highest oil content in that order.

Ideal plant type (ideotype)

1. High tillering capacity.
2. Long and broad leaves which stay green for longer period.
3. High oil content and its quality (high citronellal content).

Breeding Strategies

The varietal performance largely depends upon its genetic potential besides its interaction with the prevailing environments. Therefore, genetic amelioration keeping in view the need of the cultivators under specific situations and utilizing industry for certain specific trait. Hence, breeders have to tailor specific varieties for specific situations.

Keeping in view the allogamous nature (cross pollination) and vegetative propagation of these crops, the following breeding approaches can be applied for improvement of high oil yield and oil quality.

i. **Genetic collection and evaluation:** Collection of seeds/planting materials from its habitat and different areas of cultivation make the plant genetic resource (PGR) a strong base for studies on variability, heritability, correlation, genetic advance and plant ideotype. For any crop improvement progamme, a wider genetic base is a pre-requisite.

Therefore, widening of genetic base through introduction and collections can be instrumented in bringing new vistas in the production of high oil yielding and oil quality varieties, besides resistance to diseases and insect-pests. There is no use of exploration if the collection is preserved in perpetuity and not properly evaluated and utilized in the crop improvement programme.

ii. **Clonal selection:** Clonal selection is the only beneficially utilized method due to vegetative propagation, as there is no further segregation. Therefore, desirable clones are selected for a particular or combination of required characters. Development of "Jalpallavi" variety of citronella is another example for water stagnation tolerant variety from population of Jorlab-2 variety (Patra *et al.*, 2000).

iii. **Hybridization (intra- and inter-specific):** Keeping in view the viable and profuse seed setting, the technique of hybridization among improved and diversified clones/types is important to generate diversity. Similarly, wide crosses are also desirable to transfer resistance against biotic and abiotic stresses from related species. The inbred lines developed by continuous selfing can be crossed among themselves after evaluating for general combining ability (GCA) and the heterosis (hybrid vigour) thus obtained can be exploited by taking advantage of vegetative propagation in these crops.

iv. **Induced mutagenesis:** Development of useful mutants through gamma rays irradiation and chemical mutagens have been reported by Chaudhary *et. al.*, 1981 and Kak *et al.,* 1982 in *C. nardus*; and Kak and Kaul, 1987 in *C. winterianus*.

v. **Polyploidy breeding:** Keeping in view of the somatic chromosomes from diploid to hexaploid in the genus *Cymbopogon*, and importance of vegetative parts, efforts should be made to crate polyploids in crop by using colchicine and by other means to utilize variants thus created by doubling the chromosome number or by combining chromosome sets from different species in their amphidiploids.

vi. **Population improvement:** Recurrent selection is very useful method for population improvement and it is practiced in a synthetic variety or composite population to improve yield and quality by concentration of additive variances.

Biotechnological approaches

i. **Somaclonal variation:** Somaclonal variants generated through tissue culture led to the development of a new culture "Bio-13" of Java citronella at CIMAP, Lucknow.

Varietal Achievements

The following crop varieties of Java citronella grass have been developed from various institutes utilizing different breeding approaches.

Variety	Name of Institute	Characterization
Jorlab-2	RRL, Jorhat (Assam)	-
Mandakini	CIMAP, Lucknow	Citronellal (38%)
Manjusha	CIMAP, Lucknow	Citronellal (43%)
Bio-13	CIMAP, Lucknow	Citronellal (38%)
Manjari	CIMAP, Lucknow	Citronellal (39%)
Jalpallavi	CIMAP, Pantnagar	Citronellal
Medini	CIMAP, Pantnagar	Citronellal
Burmese	-	Citronellal (50-60%)

Stage of harvesting

The crop should be harvested when the tip of the leaves start drying and dipping down in every cut/harvest over the years.

References

Babu, C.N. 1936. Curr. Sci., 4: 739.

Kulkarni, R.N., K. Rajagopal and S. Ramesh. 1986. Plant Breeding, 97: 78-81.

Lavania, U.C. 1988. Cytologia, 53: 517-524.

Nair, R.V., N. Singh, V. Singh and K.S. Anathn. 1984. Indian Perfumer, 28: 129-132.

Pareek, S.K., D.B. Sexena, M.L. Maheshwari and R. Gupta. 1981b. Indian Perfumer, 25(3&4): 64-70.

Pareek, S.K., M.L. Maheshwari and R. Gupta. 1981a. Indian Fmg., 36(4): 37-40.

Pillay, K.P. 1961. Farm Bulletin No. 16, Ministry of Food and Agriculture, New Delhi, p. 26.

Punia, M.S., P.K. Verma and G.D. Sharma. 1986. Indian Perfumer, 30(1): 240-246.

Rao, B.L. and S.N. Sobti. 1987. Indian Perfumer, 31(1): 32.

Sharma, J.R., A.A. Naqui, R.K. Lal and H.O. Mishra. 1988. Genetica Agraria, 42(1): 13-23.

Verma, P.K., G.D. Sharma, M.S. Punia, J.S. Hooda and G. Talwar. 1990. J. Econ. Bot. and Phytochem., 1: 7-10.

Verma, P.K., M.K. Deen, S.N. Gupta and C.S. Tyagi. 2001. Genotypic variation for oil content and its quality in lemongrass germplasm. Indian J. Plant Genet. Resources, 14: 173-174.

Verma, P.K., M.L. Saini, O.P. Yadav, C.S. Tyagi and V.K. Madan. 2003. In: Proceedings of 1st National Interactive Meet on Medicinal and Aromatic Plants, CIMAP, Lucknow, pp. 114-118.

Verma, P.K., Ravinder Singh, S.N. Gupta and C.S. Tyagi. 2005. In: Course compendium of Winter School on "Advances in Medicinal, Aromatic and Under-utilized Plants Research" held at CCSHAU, Hisar from Sept. 29 - Oct.19, 2005 sponsored by ICAR.

Yadav, O.P., P.K. Verma, V.K. Madan and C.S. Tyagi. 2002. Symposium on "Perspective in Medicinal and Aromatic plants- Research and Development" held at CCS HAU, Hisar on March 7-8, 2002, pp. 33-36.

Yadav, O.P., I.S. Yadav, V.K. Madan, P.K. Verma and M.K. Deen. 2012. Abstracted in Proceedings of Seminar on "New Perspectives in Aromatic and Medicinal Plants" held at CCS HAU, Hisar on Feb. 8-9, 2012, p.146.

Suggested questions

1. What are the scientific names of two types of citronella grass?
2. Write the family to which citronella grass belongs.
3. Name 3-4 released varieties of citronella grass and variety developed through somaclonal variation?
4. What is the major nature of propagation?
5. What are the major chemical constituents of essential oil?
6. What should be the true plant ideotype?
7. Write different uses of citronella essential oil.

12

Davana (*Artemisia pallens* Wall)

Introduction

Davana (*Artemisia pallens* Wall.) is one of the important aromatic herb found abundantly in humid habitats in the plains all over India. It is popularly known as sage brush, worm wood and mug word. The davana (*Artemisia pallens*) is a small to medium height; annual and herbaceous aromatic plant. This herb is erect, and grow up to 60 cm height. Its leaves are highly divided and bears small size yellow coloured flower. Its leaves and stem found protected with greyish-white tomentum. The leaves are lobed, alternate, and petiolate.

Origin & Distribution

It is native to southern India, where it is traditionally grown in Karnataka, Maharashtra, Kerala, T.N., and A.P, growing widely in tropical to sub-tropical plains as *Rabi* season crop. It also found growing in wild form in the Himalayan region in the Simla, Kashmir valley, and Nainital hills. It is widely used by Ayurvedic healers to balance the three energies. It is an erect growing plant up to 60 cm height. It does not grow wild, but grown as a garden plant in South India. It is a crop of irrigated conditions; and do well when planted in November-December, yielding significantly higher oil yield in comparison to crop planting in July-August. The crop should be harvested 2-3 inches above ground level during bright sunny days at 50 percent flowering stage to have high oil content and oil yield.

Uses

Some of the species are valued for their essential oil with fruity smell used in perfumery and medicines. The flowers and leaves possess an oil having delicate aroma and utilised in the manufacturing extraordinary quality cosmetics and perfumes. It oil imparts a sweet, refreshingly pleasant odour on dilution. Its oil is utilised for flavouring pastries, cakes, beverages and tobacco products. It has been wide significance in Indian folk medicine against the diabetes mellitus. It is recognised with anti-pyretic, anti-helminthic and tonic properties. The oil possesses anti-fungal, anti-bacterial, anti-spasmodic, anti-nociceptive, wound healing activity, anti-diabetic, anti-cancer, anti-oxidant, anti-apasmolytic and stimulant properties. Its flowers also used for making bouquets, garlands, and spiritual gifts in most part of the country.

Floral biology

The flowers are grouped in an inflorescence, known as capitulate and bearing some pistillate ray florets located on margin and in the centre bisexual disc florets were present. Flowers are simple, sessile or peduncle, axillary or forming lax racemes, and heterogamous. The outer florets are glabrous except for a few cottony hairs, tubular, generally three-lobed. Its stigma is bi-lobed and hardly three-lobed. The inner florets are bisexual, tubular, glabrous except for a few cottony hairs, and five-lobed. It has the five, epipetalous dithecous filaments stamens which are found in free state and it has syngenecious, connective, bifid, prolonged, tapering style anthers.

It belongs to family Asteraceae (Compositae). The flowers are the main cradle of essential oil manufacture in aromatic herbs. The quantity as well as the quality of oil are highly prejudiced by the flower-pattern development. Davana (*Artemisia pallens* Bess) floral sequence and floral biology have a direct relationship with the yield of essential oil and its ingredients.

The industrial extraction discovered that biomass of floral inflorescence is the chief source of oil with the maximum 0.90 % recovery at the fool bloom stage, followed by herbage (stems and leaves) biomass.

Genotypic variability

Genus *Artemisia* contains about 470 species, generally disseminated in the temperate areas of the world. In India, its 30 species are found and most of these are confined to Himalayan belt, excepting a few of them growing in the tropical and sub-tropical plains.

The following species are reported to be prevalent.

i. ***Artemisia annua*** L.: It is considered as the greatest promising medicinal herb, containing 'artemisinin' a phyto-compound which is effective against the lethal infectious disease malaria and used in world wide in the under developed as well as developed countries.

ii. **Artemisia *pallens* Wall.:** Generally, found in the temperate areas of the sphere (Rao *et al.*, 1978). It is commonly found in Nilgiri hills and used by the tribal people for various ailments. It is highly esteemed for its exquisite fragrance.

iii. ***Artemisia umbelliformis*:** Used to treat loss of appetite and digestive spams.

iv. ***Artemisia absinthium* Linn.**

v. ***Artemisia abrotanum* L.**

vi. ***Artemisia arborescens***

vii. ***Artemisia capillaris***

viii. ***Artemisia indica***

ix. ***Artemisia vulgaris***

Out of these, *Artemisia pallens* Wall. is the most favoured in the international trade of essential oils, mainly in USA. In India also, it has monopoly in production and export of its oil.

Cytogenetics

The present day important species *Artemisia pallens* Wall. is a diploid species with chromosome numbers as 2n=16. Meiotic studies revealed that chromosomal behaviour observed during diakinesis stage was abnormal with 7 bivalents and 2 univalent, which may lead to low pollen fertility and seed set in the species.

The *species Artemisia annua* L. is also diploid but with $2n = 18$ chromosomes.

Artemisia annua

Crop Improvement

Wallaart *et al.*, (2000) worked on the biosynthetic pathway exploition of artemisinin and its analogues and enrichment of bio-active chemicals yield in plant, and advantageously exploited the genetic basis of the enzymatic process.

Keeping the importance of bio-molecule of this plant in view, and the only source of its production, the present study was carried out to systematically breed in *A. annua* genotypes with improved production of artemisinin and accumulation in the tissues of herb. Therefore, the selection trial was conducted up to four years using single plant progeny selection generation after generation. In each generation, the single plant having significantly superior artemisinin content over the previous generation was nominated and breed in polycross method for the genetic enhancement.

Mutation breeding

The flowers of Davana are tiny and seems difficult to produce genetic variability through genetic recombination. Therefore, mutation breeding techniques are more anticipated to generate genetic variability. DMS and EMS are some mutagenic chemicals belongs to the alkylating group and found common in use in crop improvement because these chemicals have potential to create more mutations as well as low chromosome aberrations.

A mutation creation experiment was conceded in Davana by using chemical mutagens for yield, morphology and quality characters during 2021 at Plantation, Spices, Medicinal and Aromatic Crops, Department of College of Horticulture, Bagalkot.

Two varities of Davana *viz.*, Theni (Cv1) and Hosur (Cv2) were evaluated for development, yield and quality constraints under the influence of chemical mutagens. DMS (0.025, 0.050, 0.075, 0.100, 0.125%) and EMS (0.01, 0.05, 0.10, 0.15, 0.20%) were applied on the plants. The growth parameters like germination (73.50%), days to first flowering (62.25 days) and 50% flowering (91.50 days), number of inflorescence (74.90) and number of capitula per plant (286.61) were observed better under EMS treatment. The other parameters were plant survival (76.56%), plant tallness (37.82 cm), number of branches/plant (23.38), head diameter (0.64 cm), 100 head weight (4.72 g), fresh herbage yield (23.01 t/ha) and dry herbage yield (5.99 t/ha) were better under the DMS treatment irrespective of Davana varieties.

The dose of 0.05% EMS and 0.050% DMS provided superior results for growth and herbage yield (stem and leaves) of Theni cultivar (Cv1), while 0.10% EMS and 0.075% DMS gives better result for Hosur cultivar (Cv2). Thus, as the application of both the mutagens increased herbage, but floral yield was reduced. Among the mutagens used in this study, DMS was more and among the cultivars, Hosur exhibited better results to the mutagens comparatively.

Varietal development

Variety	Fresh herb Yield	Oil quality
CIM–Arogya	450 -475 q/ha;	Artemisinin 0.8 -1.0 %;
CIM-Sanjivani	-	Artemisinin 1.2 %

Chemical composition of oil

Its oil has 65% esters, 20% hydrocarbons, and 15% oxygenated compounds. Oil esters are the main ingredients for its distinguishing smell. After Saponification of oil, cinnamic acid (10%) formed, while alcohol fragment provides viscous oil having an extraordinary boiling point. A novel sesquiterpene ketone known as cis-davanone in the oil is liable for its typical odour. The other isolated constituents of oil are: sesquiterpene ketone named 'artemone', novel sesquiterpenoids named.'davanafurans' and another ketone named 'isodavanone'. The different plant parts have variation in oil content, i.e. 0.53% in flower head, 0.14% in leaves including stem and 0.29% in complete plant on air dried basis.

Artemisinin is a sesquiterpene lactone alongwith endo-peroxide group, accountable for its biochemical activities and is also an exceptional natural product. Additionally, the improvement in its activities by exhausting its products is giving a novel way to the promising research at 3 levels *i.e.*, plant biology, bio-chemistry, and pharmacy. The bio-procedures the basis for workable manufacture of artemisinin or related phyto-molecules comparable to arteannuin-B and dihydro-artemisinic acid (DHA) etc., through cultivation (Khanuja *et al.*, 2008, Paul *et al.*, 2010).

In davana, Cis-davanone is the chief chemical component of essential oil and it was noted maximum at bud stage (66.36±0.60%), trailed by blooming (60.56±0.20%) and seed setting (40.54±0.80%) stage. Above findings of davana to be utilised for optimization of harvesting time to get better quality with high quantity of essential oil.

The physico-chemical characteristics of davana oil is as below (Sastry,1946).

Physico-chemical characteristics	Value
Specific gravity	0.9833 (30°C)
Refractive index	1.4880 (30°C)
Optical rotation	+ 35°
Acid value	2.4
Ester value	52°
Solubility (80 % alcohol)	Not always in 10 vol. of 70 % alcohol

Production Techonology

Davana is grown for essential oil during the winter season (first week of November). Heavy rains and high temperature at flowering imparts negative affect on growth and oil content and ultimately the oil yield. This aromatic herb thrives well on numerous kinds of soil ranging from sandy loam to medium black soils. But, well drained, organic matter rich productive sandy loam soils are found best suited. It is generally grown in Kerala, Karnataka, A. P. and T. N. Keeping the demand of this herb in view, we have a big opportunity to start its commercial cultivation in non-traditional regions to encounter the requirements export. Day-by-day, its oil demand is increasing up gradually in the international market and in India, we are not producing the oil as per the growing demands. Therefore, necessity has been realized in relation the development of new varieties along with production technology of this crop.

Growers usually unable to prepare their own seed due to harvesting of crop before seed set, for both ornamental and essential oil purposes. That's why farmers depend on seed supplying agencies for seed to grow the next season crop (Farooqi *et al.*, 1990).

Yield and Oil Content

The fresh herbage yield of davana, including inflorescence, is 12-13 t/ha. However, in laboratory, oil content in the whole plant after after shade drying for two days is 0.29%. On average basis, recovery of oil is about 0.2% in large-scale distillation unit, and oil yield is 12-13 kg/ha. When oil is extracted from the entire plant, the flower-heads is the main contributor of the oil. The percent input from stems and leaves in the oil is very little.

Suggested questions

1. What are the two prevalent species of Davana?
2. Write the family to which Davana belongs.
3. Name the released varieties.
4. What is the ploidy level and somatic chromosome number of two species of Davana?
5. What are the major chemical constituents of essential oil?
6. What are the effective chemical mutagens in Davana?
7. What are the different uses of Davana essential oil?

13

Dill Seed/Sowa (*Anethum graveolens* L.)

Introduction

Dill seed/sowa belong to the genus *Anethum* and it is derivative of Greek word 'aneeson or aneeton', meaning thereby strong smell. The botanical name of dill seed is *Anethum graveolens* L. It is the only species falling in the genus *Anethum*, and known as East Indian dill or *Sowa* (*Anethum graveolens* var. *sowa* Roxb. ex, Flem.) in India. It is cultivated as a foliage crop in *rabi* season stretched in the Indian sub-continent.

Origin

It seems to be originated in South-east Europe or South-west Asia and also native to Central Asia, Mediterranean and southern USSR. It belongs to India, as it is found growing in the states of Maharashtra, Gujarat, A.P., M.P. and Rajasthan.

Uses

- Used as in abdominal discomfort, colic and promoting digestion.
- Cures 'vata', 'kapha', ulcers, eye diseases and uterine pains.
- Paste of dill seed with milk is used for external use in rheumatic and swellings of joints.
- Attributed to tonic, rejuvenating and intellect promoting properties.
- Used as condiment.
- Seeds are used as carminative, aromatic, galactogogue, mildly diuretic, stomachic and stimulant.

- Essential oil extracted from seeds relieves intestinal spasms and griping to settle colic.
- It also stimulates flow of milk in lactating cows and ladies.

Floral biology

The plant takes height upto 90 cm having cylindrical stems and leaves with alternate arrangement. The leaves are dissected three or four times into pinnate. The yellow flowers develop into umbels. Its fruits are compressed, oval shaped, winged about one-tenth inch wide, having three longitudinal ridges on the back and three dark lines. The seeds are smaller, flatter and lighter having pleasant aromatic odour. The crop belongs to family Apiaceae (Umbelliferae).

Classification

Earlier genus *Anethum* was included under *Peucedanum*, and further had four different species:

a) *Peucedanum angustifolium*

b) *Peucedanum graveolens* (*syn. Anethum graveolens*)

c) *Peucedanum ostruthium*

d) *Peucedanum sativum*

Presently, dill seed is grouped under genus *Anethum*, consisting of two most common species:

i. *Anethum graveolens* Linn. (European dill)

ii. *Anethum sowa* Roxb.ex. Flem. (Indian dill)

The above two are presently considered as two types or varieties. However, these differ in their oil quality having carvone and dillapiole content. European dill was introduced from UK by Regional Research Laboratory, Jammu and reported that its oil quality was comparable. Dill is rarely grown as herb in India for flavouring soups, sauces, etc., requiring only young leaves.

Cytogenetics

It is a diploid species with 2n=22 chromosomes, and genomic chromosomes x=11.

Genetic variability, heritability and correlation

Keeping in view the presence of major heterogeneity and important genetic resources in landraces, their evaluation and use in breeding programs as starting material is very important. It provides new potential to improving crop productivity and adaptability. The genetic improvement in dill seed is limited due to narrow genetic pool.

The crop is grown for both seed as well as foliage/leaf. Therefore, selection of genotypes with high leaf yield is indispensable, and depend upon high heritability and genetic advance. Hence, collection of local genotypes and assessment of their genetic variability is of great importance. The study in this regard revealed significant differences among the genotypes along with high heritability (85.5%) coupled with high genetic gain (34.7%) for leaf yield, indicating better chances for its improvement.

The phenotypic (PCV) as well as genotypic coefficients of variation (GCV) were found to be low, medium and high among different genotypes under study. The leaf yield was found to be positive and highly significantly correlated with yield and other related traits.

The genotypic correlations being higher than the phenotypic correlations between different traits revealed that simple selection for yield and their linked characters would be operative for genetic enhancement of leaf production. Therefore, high leaf yielding genotypes *viz.* ACKD-8, ACKD-6, ACKD-11, ACKD-16, ACKD-14, ACKD-28, ACKD-23, ACKD-29 and ACKD-30 were found to be promising and suggested for commercial crop cultivation.

Genetic diversity

Different genotypes of dill seed crop studied for different agro-morphological characters suggested the importance of plant height, fresh herb weight, number of leaves per plant, diameter and the number of umbels per plant. Due to its multiple medicinal and culinary uses, it is constituted as a vital substitute crop with high added value for small scale growers.

The assessment of phenotypic diversity of Greek dill landraces further revealed that crop productivity associated traits contributed heterogeneity, which could be potentially exploited by the breeders for improving seed production.

Cluster analysis classified different genotypes into four groups determining evolutionary relationships and classification position of genotypes from different populations for selecting the appropriate genotypes for future crossing programme with the purpose of creating diversity.

Chemical composition of Dill seed oil

Major constituents	Indian dill seed (%)	European dill seed (%)	Reference
Carvone content	20.7	34.5	**Baslas *et al.*, 1971**
Dihydrocarvone	14.3	12.5	
D-Limonene	21.4	10.0	
D-phellandrene	11.4	6.0	

Varietal development

The following varieties have been developed:

Variety	Suitable conditions	Characters	References
Gribovskii	Early sown	High herbage & seed yield	Mukhanova, 1975
Lesnogorodskii	Late sown	High oil and seed yield	Sychev and Shmanaeva, 1989
Dura	Normal sown	High herb yield of good flavour	Halva *et al.*, 1987
K-299 & K-253	Normal sown	High oil yield	Kuryanchik, 1976
Mammoth Long Island (MLI)	-	High quality oil	Chubey and Dorrell, 1976
Haldwani Selection	-	High seed yield	Gupta, 1971
Stamm Weibull Herkules	-	High oil content at early stage High oil content & quality	Krodel, 1967

Production technology

- Perform well in light sandy soils with full sunlight.
- Can tolerate soils pH up-to range of 5.3 to 7.8.
- *Kharif* season crop requiring warm to hot summers.
- Seed rate of 5 kg/ha is optimum for sowing by drill at 30 cm row spacing.
- October sown crop was also found to give the highest yield as well as oil content.

Suggested questions

1. What are the two prevalent species of Dill seed?
2. Write the family to which Dill seed belongs.
3. Name the 2-3 released varieties.
4. What is the ploidy level and somatic chromosome number?
5. What are the major chemical constituents of essential oil?
6. What are the different uses of Dill seed essential oil?

14

Geranium (*Pelargonium graveolens* L. Herit)

Introduction

Geranium is also known as cranesbill, scented geranium/rose geranium. It is adaptable over a wide range of soil conditions and drought tolerant. Being perennial aromatic herb, it is cultivated for rose scented essential oil extracted from leaves and tender shoots. The oil thus obtained is also called pelargonium oil or rose-geranium oil. The oil is of varied colour from colourless to pale yellow-brown or greenish. The genus *Pelargonium* belongs to the family Geraniaceae.

Origin

Geranium is native to sub-tropical southern Africa, and the most popular of bedding and green-house plants. The genus *Pelargonium* consists of about 600 species, which are annual, biennial, and perennials found in Cape province of South Africa.

Botany

The cultivated geraniums may belong to herbaceous to woody with thick fleshy leaves of round to lobed shape and deep cut. The flowers are present in terminal clusters with varying colours from white to pink, deep red and violet. Cultivated species are hybrids, having double flowers of mixed colours. Species are mostly propagated by stem cuttings, producing flowering plants in 4-5 months. Geraniums can also be multiplied from seeds, resulting in breeding and selection of useful lines.

Genotypic variation

Geraniums are of various categories as given below:

- Geraniums (*P. domesticum*) are largely derived from *P. cucullatum, P. angulosum,* and *P. grandiflorum,* having large pansy flowers.
- Bedding geraniums (*P. hortorum*) are complex hybrid, largely derived from *P. inguinans* and *P. zonale*, familiar to garden culture.
- Ivy/hanging geraniums (*P. peltatum, P. lateripes*) are grown as basket plants for indoors and outdoors.
- The aromatic/scented-leaved geraniums include *P. abrotanifolium, P. capitatum, P. citrosum, P. crispum, P. graveolens, P. quercifolium, P. radula* and *P. odoratissimum*.

Uses

- African *Pelargonium* are commercially important for essential oil used in perfumery.
- Oils are also used in soaps, ointments, tooth paste and dusting powder industries.

Cytogenetics

Diploid chromosome numbers as 16,18, 20, 22, 36, 45, 60, 72, 81, 88, and 90 are reported from 37 species of genus *Pelargonium* with genome number as x = 8, 9, 10, and 11 (Fedorov, 1974).

Species	Diploid chromosomes (2n with x = 11)	References
P. crispum	22 (diploid)	Tokumasu, 1974
P. tomentosum	44 (tetraploid)	Demarne and Vander Walt, 1990
P. radens	88 (octaploid)	Demarne, 1989
P. graveolens	88 (octaploid)	-do-
P. vitifolium	88 (octaploid)	-do-
P. capitatum	66 (hexaploid)	-do-
P. graveolens	66 (hexaploid)	Angadi, 1990
P. graveolens	144	Janick *et al.*, 1977
P. crispum	22 (Diploid)	Tokumasu, 1974

It is obvious that basic chromosome number x=11, and species are either tetraploids, hexaploids or octaploids. However, variant chromosome number with n=72 has also been reported for *P. graveolens* with a tissue culture induced tetraploid (2n=144).

Breeding methodology

Breeding for crop improvement is based on creation of genetically variable populations from which superior genotypes are selected or isolated. For that purpose, traditional breeding depends on sexual cycle as variation and released. But it may impose a restriction on genetic improvement of a polyploidy, complete sterility and vegetative propagated crop, like geranium, where seed is not the economic part. Thus variation occurred by spontaneous mutations are inevitable and becomes necessary. Therefore, natural variation available in the country needs to be collected and evaluated first, before attempting mutation breeding for creating induced mutants.

i. **Germplasm collection and evaluation:** Such efforts were initiated under All India Coordinated Research Programme on Medicinal &Aromatic Plants at Horticultural Research Station, Kodaikanal and Indian Institute of Horticultural Research, Bangalore. At Kodaikanal, a clone PG-7 was isolated with high herb yield potential and high essential oil content, having highest refractive index, specific gravity and geraniol (Vadivel and Sampath, 1981). Further, single plant selections resulted in a selection resistant to tip rot (caused by heavy rainfall conditions) and very high herbage yield by Cinchona Department (Doreswamy and Sundaram, 1978). Later on, two promising strains were selected through clonal selections by IIHR, Bangalore (Kumar, 1986 and Angadi, 1990).

ii. **Hybridization:** Yoshida (1969) reported to obtain a hybrid between *P. roseum* and *P. denticulatum* to have a high yield of good quality oil.

   Angadi and Kumar (1989c) attempted hybridization between Reunion mutant (Reunion pubescent) and Alg-4n to combine the good quality of the former with better oil content of the latter. The hybrid exhibited heterosis and was fully fertile. It was further evaluated for oil content and oil quality.

iii. **Mutation breeding:** It has a special advantage of adding specific characters to a vegetative propagated crop of geranium for higher herbage, oil content and oil quality; where varieties are highly heterozygous clones, not suited for improvement by selection of recombinant due to inherent sterility following hybridization or improvement by back-crossing.

Excised leaves of both Algerian and Reunion types were gamma-irradiated (Co-60) and propagated by adventitious bud technique to overcome the problem of low recovery of mutants due to chimera formation and diplontic selection in irradiated cuttings. Variants for morphological traits and slightly higher oil content were also isolated and evaluated for resistance to wilt.

Varietal development

Two types of geranium as given below are under cultivation in India.

a) **Algerian/Tunisian type:** Grown in the Nilgiri hills in large areas, having slender dark pink flowers. This type yields 50-60 percent more oil than reunion type. These are suitable for dry conditions.

b) **Bourbon/Reunion type:** Grown in the Nilgiri and Annamalai hills. This type is more sturdy with light pink flowers and more suitable for wet conditions; and preferred for cultivation in plains. The oil content is higher during summer season and only terminal leaves contain more oil than middle and basal ones.

After evaluation of these two types at HRS, Kodaikanal; CIMAP, Kodaikanal; and IIHR, Bangalore, Algerian types were found more superior than the Reunion type, having higher herbage yield, oil content and tolerance to wilt disease; and rejuvenation capacity after heavy pruning of one-year old plants.

HRS, Kodaikanal reported that Reunion type contained superior quality oil with 59.43% citronellol; followed by Algerian type clone PG-7 with low oil content but high citronellol content (57.60 %), whereas clone PG-20 had high oil content (0.3%) but lacked in geraniol fraction.

Following varieties have been developed so far

1. **CIM-Pawan:** It was developed and released from CIMAP, Lucknow (U.P.), yielding essential oil to the tune of 50 kg/ha with 21 percent geraniol content.
2. **KKL-1:** Developed at Horticultural Research Station, Kodaikanal.
3. **Narmada:** It is a rose scented geranium developed as soma-clonal variant of Indian Cultivar **Bourbon** developed at CSIR-CIMAP Research Centre at Banglore.
4. **Sel-8(Bourbon/Reunion Type):** Developed at IIHR, Bangalore.
5. **Kelkar & Ooty:** Developed at IIHR, Banglore
6. **Hemanti (Algerian/Tunisian type):** Developed at CIMAP, Lucknow is rich in citronellol and suitable for cultivation in the plains of north India.
7. **Bipuli (Bourbon/Reunion type):** Developed at CIMAP, Lucknow is rich in both geraniol and citronellol content.
8. **Kunti:** Developed at CIMAP, Lucknow

Chemical composition of different species

***Palergonium* sp.**	**Chemical compositions**
P. graveolens	Isomenthone (mint scent), negligible citronellol & geraniol
P. radens	Isomenthone (mint scent), negligible citronellol & geraniol
P. capitatum	Mildly rose scented (Geraniol upto 2.82%)
P graveolens x P. radens	Both citronellol and geraniol
P. graveolens x P capitum	Both citronellol and geraniol

References

Angadi, S.P. 1990. Annual Progress Report of the project. IIHR, Bangalore.

Angadi, S.P. and Vasantha Kumar, T. 1989c. In: National Symposium on "Recent Advances in Genetics and Plant Breeding in India" held at B.H.U., Varanasi from Nov. 15-16, 1989.

Demarne, F.E. 1989. Genetic improvement of geranium rose (Pelargonium sp.): Systematically karyological and biochemical contribution. D. Sc. Thesis, University of Paris.

Demarne, F.E. and Vander Walt., J.J.A. 1990. South African of Plant Soil, 7(1): 36-39.

Doreswamy, K. and Sundaram, M. 1978. Indian Perfumer, 22 (1): 43-45.

Fedorov, 1974. Chromosome numbers of flowering plants. Otto Koeltz Science Publishers, Germany.

Janick, J., Skirvin, R. M. and Janders R.B. 1977. J. Hered., 68: 62-64.

Skirvin, R.M. and Janick, J. 1976. J. Amer. Soc. Hort. Sci., 101 (3): 281-290.

Tokumasu, S. and Kato, M. 1979. Euphytica, 28: 329-338.

Vadivel, B. and Sampath, V. 1981. In: IV Workshop on AICRP on M&APs held from 31 Aug. - 3 Sept., Madurai.

Yoshida, T. 1969. Japan Agric. Res. Quart., 4(3); 23-26.

Suggested questions

1. What is the major species of Geranium?
2. Write the family to which Geranium belongs.
3. Name the 2-3 released varieties.
4. What is the ploidy level and somatic chromosome number of major species with genomic number?
5. What are the major chemical constituents of essential oil?
6. What are the different uses of Geranium essential oil?

15

Haldi/Turmeric (*Curcuma* species)

Introduction

Genus *Curcuma* belongs to the Zingiberaceae family, and containing of about 122 species. These are grown profusely in Asian countries including, India, China, and Southeast Asia, Northern Australia and New Guinea. Like ginger, it is also the popular herb of this family for the utilization of its underground rhizomatous stem by human beings. Botanically it is named as *Curcuma longa* L. Its bright yellow spicy powder is got from the dried rhizome is termed turmeric powder, which is the final product of this herb. In addition to this curcumin bio-active compound is also extracted by cursing rhizomes of turmeric. The plants turmeric are considered to be a hybrid and sterile leaving vegetative propagation from rhizome as the single mode of propagation. India is recognized as the power house of turmeric production. Along with us, our neighbours in South East and Eastern region, including Malaysia, Thailand, Bangladesh, Cambodia, Philippines, Indonesia and even China are considered major producers.

Floral Biology

All species of *Curcuma* genus have almost alike flowering spikes that mainly originate from the topmost region of the rhizome (pseudo-stem) or sometimes on a separate stem directly from the rhizome. These two stems differ only in the inner portion of rhizome, which differ in colour, i.e., creamy, white, orange, yellow, bluish-green, blue, and sometime black also. The rhizomes exhibit commercial value as a medicine as well as food spice. *Curcuma longa* L. (Syn. *Curcuma domestica* Val.), is the common turmeric. It is the top ranking herb based on its economic potential among all the members of this genus and cultivated in India about 150,000 hectares.

Origin

The genus *Curcuma* L. is inherent to the tropical and sub-tropical regions of Asia and Northern Australia. The true native (original birth place) region of turmeric is probably South or Southeast Asia (Vietnam, China or western India), where this rhizomatous herb is static comprehensively cultivated. *Curcuma petiolata* seems to be originated in Thailand and Malaysia, where it is found widely distributed and also named as "Jewel of Thailand" retaining to its dominant utilisation as ornamental plant and medicinal herb by the residents of Thailand.

Uses of *Curcuma domestica/Curcuma longa*

- Today, the chief use of turmeric is in form of household spice which is obtained from the rhizomes after boiling, drying and powdering.
- The turmeric is used by human kind perhaps started for its yellow colour dye, which seems to have a amusing history in colouring foods items. Now-a days, it is utilized in the several industries, for example, pharmaceuticals, processed foods, confectionery sauces, and also in textiles.
- It has a traditional yellow turmeric colour and also imparts new colours on mixing with other dyes for cotton, silk, and woollen fibres.
- The oil extracted from turmeric used for flavouring the curry and also have industrial food items application.
- Different types of turmeric preparations found in market for the people both as a medicine and food.
- Turmeric is commonly used either as a fresh rhizome or the dried rhizomes yellowish powder which is available in various packings
- Rhizome contains bio-active compounds having numerous functional activities, for example, anti-inflammation, anti-microbe, anti hyper-cholesterolemia, anti-rheumatic, antivirus, anti-fibrosis, anti venomous, anti-hepatotoxicity, anti-diabetes, anti-cancer, and gastro protector.

Utility of different *Curcuma* Species

***Curcuma* species**	**Uses**
Curcuma longa, C. aromatica	Toiletry articles and medicines
C. ochrorhiza, C. kwangsiensis, C. caesia, C. zedoaria, C. pierreana,	Among the South-East Asian nations as Folk medicines
C. roscoeana, C. alismatifolia	Floricultural significance
Curcuma amada	Medicine, culinary preparations, pickles & salad
C. malabarica, C. zedoaria, C. montana, C. pseudomontana, C. decipiens, C. angustifolia, C. rubescens, C. caulina, C. haritha	Arrowroot manufacturing

Genotypic variability

Sufficient genetic variability has been reported by several workers in the genus *Curcuma*. Now-a-days, the diverse plant species belonging to the genus *Curcuma* are getting more attention at global level. Very important notable species of *Curcuma* reported by different researchers, are: *Curcuma aromatica*, *Curcuma longa*, *Curcuma amada Roxb*, *Curcuma caesia Roxb*, *Curcuma purpurascens*, *Curcuma manga*, *Curcuma zedoria* , and *Curcuma xanthorriza*.

Curcuma longa (Desi haldi)

Curcuma amada (Ama haldi)

Curcuma caesia (Kali haldi)

Cytogenetics

Curcuma, a rhizomatous, perennial and herbaceous group of plant. It reflects enormous genetic diversity in ploidy levels, which is self-evident from its cytological findings, wherein different chromosome numbers of 2n = 22, 42, 63, 77, 105 etc., have been reported by different workers while working different plant samples in different regions of the world.

Moreover, nonstop dispute regarding the basic chromosome number in *Curcuma* (x=7, 8, 16 and 21) have been recounted in various cytological investigations by the different workers. The significant evidence related to the somatic chromosome number is found reported for the genus *Curcuma*.

The species were divided into two groups as below:

Group I: 2n = 42 (*C. comosa*, *C. haritha*, *C. manga*, *C. motana*)

Group II: 2n = 63 (*C. caesia*, *C. longa* and *C. sylvatica*).

Group I: Exhibited the prevalence of bivalents, however occasional quadri-valents were also encountered. Almost 48% of the PMCs examined in *C. manga* exhibited 21 bivalents (II) meiotic configurations, 16% in *C. haritha* and 32% in *C. comosa*.

Group II: As expected showed the presence of trivalents, besides bivalents, univalents and quadri-valents. Around 32% of the PMCs examined in *C. sylvatica,* exhibited 21 trivalents (III) meiotic configurations, 8% in *C. caesia* and 24% in *C. longa*. Overall, low occurrence of multivalent associations as compared to bivalents showed that *Curcuma* is an allopolyploid complex.

Moreover, the basic chromosome number, x=21 seems to be very high. Therefore, it is suggested that the genus *Curcuma* come in existence through hybridization amog the species having diverse chromosome numbers of 2n = 24 and 18, resultant evolved a dibasic amphidiploid species. Further, it is also described in literature that 25 species of *Curcuma* possess 2n=42 chromosomes.

Cytogenetics of different species under genus *Curcuma*

Species	**Diploid chromosomes (2*n*)**
C. aeruginosa Roxb.	63, 84
C. alismatifolia	32
C. amada Roxb.	42
C. angustifolia Roxb.	42
C. amarissima Roscoe	63
C. aromatica Salisb.	63
C. aurantiaca Zijp	42
C. attenuata Wall.	84
C. brog Valeton	63, 64, 70
C. candida	42
C. caesia Roxb.	63
C. colorata Valeton	62, 63
C. comosa Craib	42
C. comosa Roxb.	63
C. decipiens Dalzell	42
C. elata Roxb.	63
C. haritha Mangaly	42
C. heyneana Valeton	63
C. latifolia Roscoe	63
C. longa L.	63, 48
C. malabarica Velay.	42
C. mangga Valeton	63
C. eilgherrensis Wight	42
C. oligantha Trimen	40, 42
C. petiolata Roxb.	42
C. phaeocaulis Valeton	62, 63, 64
C. purpurascens Blume	63
C. raktakanta Mangaly	42, 63
C. rubescens Roxb.	63

C. roscoeana Wall.	42
C. soloensis Valeton	63
C. sessilis Gage	46, 84, 92
C. thorelii Gagnep.	34, 36
C. viridiflora Roxb.	42
C. wenyujin Y.H.	63
C. zedoaria (Christm.) Roscoe	63
C. zanthorrhiza Roxb.	63
C. alismatifolia Gagnep.	32
C. gracillima Gagnep.	24, 32
C. gracillima Gagnep.	40
C. harmandii Gagnep.	20
C. parviflora Wall.	24, 28, 34, 30, 32, 42, 36, 56
C. rhabdota Sirirugsa	24
C. thorelii Gagnep.	34, 36

Breeding methodology

Many attempts for genetic improvements have been reported in *C. longa* and only a few attempts reported in *C. amada*. Till today, in India nearly 20 high yielding cultivar of *C. longa* and one in *C. amada* have been developed through clonal selection from germplasm, mutation breeding or open-pollinated progeny (true turmeric seedlings) selection.

Curcuma species have significant intra- and inter-specific difference for plant morphological traits as well as for the biologically active principles. In addition to above, floral spike variation as well as the under-ground rhizome structures along with composition of curcumin, oleoresin and essential oil also found variable.

The research work for morphological characterization for the diverse species/ varieties of *Curcuma* has been carried by several researchers at different locations. Recently, its characterization at molecular is also started, except for some genetic fidelity of micro-propagated plants and isozyme-based description.

Chemical composition of oil

Turmeric rhizomes contain 4-6 percent essential oil. This genus was inspected from the oil components point of view. The common phyto-constituents originate in *haldi* are flavonoids, phenols, terpenoids, alkaloids, tannins, steroids, glycosides and saponins. The other important bio-components belongs to *Curcuma*, are curcuminoid and xanthorrizhol. These are very important compounds exhibiting very similar bio-activities and important for health.

Varietal achievements

A large number of varieties have been developed by different institutes/ universities for different agro-climatic conditions. Some of them are as below:

Variety (Year)	Institute/Univ.	Area/State
Hisar Kanchan	CCS HAU Hisar	Haryana
Suvarna	IISR, Kerala	Odisa, TN, HP, AP, Kerala
Sugma	-do-	-do-
Sudarsana	-do-	-do-
Prabha	-do-	-do-
Pragati	-do-	-do-
Surama	-do-	-do-
Roma	-do-	-do-
Allepy Supreme	-do-	-do-
Kedaram	-do-	-do-
Rasmi	-do-	-do-
Suranjana		
Punjab Haldi-1	PAU, Ludhiana	Punjab
Punjab Haldi-2	-do-	-do-
Sona		Kerala
Varna		Kerala
Phule Swarupa (2006)		-do-
GNT-2 (2018)	GAU, Gujarat	South Gujarat
CO-2 (2016)		TN, Karnataka, Maharastra, AP, TelangaSna
TCP-64	UBKV, WB	WB
Termeric-3 (2021)	-do-	WB, Bihar, TN
Narendra Haldi-1	NDUA&T, UP	UP

Stage of harvesting

The crop is harvested when aerial/upper groung portion having leaves becomes dry and its colour turns light brown to yellowish. The crop is harvested during winter season after 7-9 months depending upon the sowing time.

Suggested questions

1. What are thescientific names of three species of Turmeric?
2. Write the family to which Turmeric belongs.
3. Name the 3-4 released varieties of Haldi.
4. What is the ploidy level and somatic chromosome number of major species with genomic number?
5. What are the major chemical constituents of essential oil?
6. What are the different uses of Desi Haldi?

16

Jasmine (*Jasminum grandiflorum* Linn.)

Introduction

Jasmine is commonly known as Chameli, Champa, Juhi, Mogra, and Motia in India. French jasmine, Spanish jasmine and Catalonian jasmine are also found in India. It consisted of both as vines/climbers and shrubs.

Origin

'Jasminum' word is derived from an Arabic word 'Yasmin.' It is native to the Himalayas and mostly found in tropical and sub-tropical regions. It originated in the north-east foothills of Himalayas in Indo-Burma. It is known for characteristic smell and beautiful colours of flowers for our indoor as well as outdoor gardening. Jasmines are a group of ornamental domesticated plants, grown perennial for their fragrant flowers. However, its potential for production of perfume has also been first exploited in France.

Distribution

It was first introduced on commercial scale for production of essential oil in Egypt, Morocco, South Africa, Syria, Algeria, Tunisia, United Arab Republic, China and India. Keeping in view the manual picking of flowers, it is labour based agro-industry. Presently, its flowers and essential oil is exported to Europe, Russia, Japan and USA.

In India, the first perfume based plantation industry was established in 1972 at Mettupalayam (T.N.), and subsequently spread in the surrounding states. As a result, it is now commercially grown in Maharashtra, Uttar Pradesh, West Bengal, Rajasthan, Andhra Pradesh, Karnataka and Tamil Nadu states; and now a large number of industries have been established in T.N., Karnataka, Andhra Pradesh and Maharashtra.

Genotypic variation

More than 200 varieties/species are found in sub-tropical and tropical world growing wildly. Forty-three (43) species are known to grow in different parts of India. It belongs to family Oleaceae. Efforts for genetic improvement in jasmines have been made on the following species found in India.

Jasminum sambac

Jasminum grandiflorum

Jasminum officinale

Jasminum humile

Jasminum fruticans

Jasminum mesnyi

1. ***Jasminum sambac*:** Commonly known as Arabian jasmine, Tuscan jasmine, Bela and Mogra. Buds and beautiful white colour flowers with sweet scented smell are frequently used for adorning of hairs, preparing garlands, God/ Goddess worshipping and decoration purposes. Temperate climate is congenial for its prosperous well. It is a perennial vine having flowers of multiple layers and found growing all the year round in India.
2. ***Jasminum auriculatum*:** Locally known as Juhi, and buds and flowers are used for making garlands, adorning hairs, worshipping of God & Goddess and decorative purposes.
3. ***Jasminum grandiflorum*:** Generally known as French jasmine, Chameli, and Royal jasmine. It is chief source of essential oil extracted from flowers. Besides, leaves have been reported to be used as herbal medicine. The white flowers are beautiful with a characteristic fragrance and used for ornamental purposes.. It is a deciduous shrub in nature, grown in temperate to sub-tropical climatic conditions.
4. ***Jasminum pubescence* (*Syn J. multiflorum*):** Downy jasmine, Kundamu, Kundphul, and Kundum are its other names. It is the hardiest jasmines and hence least affected by pests and diseases. The flowers are non-scented, but very attractive in appearance.

5. **Common Jasmine (*Jasminum officinale*):** Generally recognised as Summer jasmine, True Jasmine and Poet's Jasmine. Its plants are climbers in nature and of deciduous type. The large white flowers have potent scented during summer season. The flowers are having five petals.
6. **Italian Jasmine (*Jasminum humile*):** The plants do well in temperate zone with cold climate in India. The flowers are yellow in colour and shaped like a funnel with fragrance.
7. **Pink Jasmine (*Jasminum polyanthum*):** It is an evergreen climbing type Jasmine having vigorous growth and grown for aesthetic purposes in kitchen gardens. Dark green foliage and beautiful red-pink flowers are worth seeing.
8. **Wild Jasmine (*Jasminum fruticans*):** This evergreen plant has dark green foliage with yellow flowers blooming in bunches but without any smell. Plants require less attention and even can survive under drought conditions with slow growth.
9. **Primrose Jasmine (*Jasminum mesnyi*):** It is an evergreen bushy type plant having yellow flowers and dark green foliage with vigorous growth. The excessive growth can be controlled by pruning at regular intervals. This is very much popular and favourite among people having love with landscaping.
10. **Winter Jasmine (*Jasminum nudiflorum*):** Yellow flower
11. **Dwarf Jasmine (*Jasminum parkeri*):** Yellow flower
12. **Forest Jasmine (*Jasminum abyssinicum*):** White flower
13. **Lemon scented Jasmine (*Jasminum azoricum*):** White flower

Floral biology

Generally, jasmines are climbers by habit, but can be trained as bushes also. Flowers are hermaphrodite having two bilobed anthers (androecium), and ovary (gynoecium) with one ovule, simple style and bilobed stigma. Ovary is superior, bilocular having axial placentation. Anthesis have been reported at four different times during evening hours between 5-10 P.M., and bursting of anthers (dehiscence) at three different times. Flower buds opening is maximum during 6-7 P.M., and period of blooming vary from 3-4.5 months. It belongs to family Oleaceae

Jasminum grandiflorum has been reported with floral dimorphism expressing long and short carpel. Distinct differences in flowering behaviour were observed in **pin** and **thrum** types. The long carpel type (pin) had prolific flowering, whereas short carpel type (thrum) expressed shy flowering.

Jasminum auriculatum is having flowers of both pin and thrum types. Thrum type plants were found to be superior to pin type in floral bud length, corolla tube length and diameter of open flower.

Uses

Flowers are used for various purposes and have prime importance as below:

- It has utility in Ayurveda medicines as an effective relaxant.
- Flowers are also used in cosmetic industries in making scents and perfumes.
- The Gajra made from flowers of jasmine is favourite as hair accessory among Indian women.
- Flowers are also used by health and skincare industries to prepare herbal tea and aromatherapy reducing stress and promoting healthy well-being.

Cytogenetics

Genotypes having both diploid with 2n=26 chromosomes and triploid with 2n=39 chromosomes have been reported by Sharma and Sharma (1958), and Murty and Khanna (1971) in different species. However, triploid strain **'Pink Thrum'** has also been identified as tetraploid with 2n=52 chromosomes by Karmaker and Srivastava (1987). Therefore, the basic chromosome number (genome) x=13.

Seed setting

Study conducted at IIHR, Bangalore grouped Jasmine species into 5 categories as given below:

Jasmine species	Seed setting (%)
J. auriculatum	Profuse seed setting (70-80)
J. calophyllum and J. flexile	Moderate seed setting (5-10)
J. angustifolium, J. grandiflorum, J. savissinum	Poor seed setting (1-5)
J. arborescens, J, rigidum	Very poor seed setting (< 1)
J. humile, J. nitidum, J, officinale, J. pubescens	No seed setting

Breeding methodology

1. **Clonal Selection:** It has been used in different *Jasminum* species as given below:

   a) ***Jasminum auriculatum*:** A high yielding variety known as **'Parimullai'** was evolved at TNAU, Coimbatore. It yielded 43 q/ha flower compared to 25 q/ha in ordinary types along with resistance to gall-mite infestation.

b) ***Jasminum auriculatum***: Several selections were evaluated at TNAU, Coimbatore, resulting in a selection yielding over 80.8 q/ha of flowers with 0.34% concrete. The floral buds possessed short corolla tube with bold bud size, and later releasing as variety **CO-1 'Mullai'**.

c) ***Jasminum grandiflorum***: Six clones were identified based on their distinctive morphological characters of economic importance. The clone **Jg3** Scions of the strain **'Lucknow'** was found the most promising, yielding huge flower production to the tune of 101.44 q/ ha with concrete recovery of 29.42 kg/ha. Finally, this strain was released as variety **CO-1 'Pitchi'** by TNAU, Coimbatore.

d) ***Jasminum grandiflorum***: A high yielding strain **'Pink Pin'** developed at IIHR, Bangalore with flower yield potential of 1000 q/ ha and concrete recovery of 35 kg/ha was named as **'Arka Surabhi'**.

2. Hybridization

The hybridization for genetic enhancement seems to be difficult due to seed setting problem in majority of species and their hybrids as given above. As the flowers bloom during night therefore it is very difficult task to emasculate and pollinate the flowers as style are long and delicate in some cases and having high tendency of breaking down of stigma. This problem is further increased several folds due to attracting fragrance thus frequently visited by insects. It was noticed that even very tiny insect can move up and down in the corolla tube, therefore, stigma got dust with pollen grains. Generally, bud pollination is followed to avoid the interference of insects. In jasmine fruit, normally one seed is found and the growth of new seedling is quite slow. Keeping all these points in consideration, jasmine breeders has to developed a suitable breeding technique and method to regulate the visit of the minute insects before emasculation and after pollination to achieve the fruitful results.

At IIHR, Banglore, a large number of crosses were attempted involving different *Jasminum sp. viz. J. arborescence, J. flexile, J. calophylum, J. grandiflorum, J. nitidum J. humile var. wallihianum,* and *J. rigidum.* But most of the cross combinations failed to set seed. Certain combinations produced very shrivelled seeds, except from one combination involving the parent *J. flexile* and *J. suavissimum,* while there was no seed germination in other cases.

Eighteen (18) hybrid progenies obtained from two parental combinations *viz.*, Mutant x Parimullai and Mutant x Long point were evaluated for their variability. High genetic variability were observed for plant volume and flower yield per plant. Mutant x Long point exhibited the highest flower yields consistently.

Srivastava and Shamasundaram (1988) reported that species *Jasminum gradiflorum* is highly heterogenous and presence of sterility resulted in negligible seed set to allow heterozygosity.

Srivastava and Karmakar (1986) reported that *J. grandiflorum* required darkness for seed germination.

Mutation Breeding

To create large spectrum of genetic variability in the *Jasminum* species, a range of physical and chemical mutagens were used for modification in traits of economic value. Generally, the induced genetic variability is followed by cycles of recombination and selections could result in improvement of required features, which otherwise may not respond to selection. Once a desirable mutant is achieved, it can be easily multiplied by vegetative means.

In *Jasminum auriculatum,* it was reported to obtain a spontaneous mutant with larger flowers, which excelled in length and width of floral bud, length of corolla tube, diameter of open flower, number of petals & length and width of petal and 100 floral buds weight in comparison to the source plant.

Chezhiyan *et al*. working at TNAU, Coimbatore reported a variegated mutant in *Jasminum auriculatum* by seed irradiation with 7.5 kR, 12.5 kR, 30 kR and 10 kR + Ethyl methane sulphonate (2 %) doses. Normally, the plants come in flowering in 12 months, but due to short in height and having variegated type foliage, this mutant plant achieved the ornamental impotance.

Nambisan *et al*. (1980) obtained two dwarf and resistant to leaf spot disease (*Cercospora jasminicola*) through induced mutantations in *Jasminum grandiflorum*, but both of them were lower in concrete percentage.

Kumar *et al*. reported sensitiveness of *J. grandiflorum* cuttings to gamma radiations. Softwood cuttings of the variety **'CO-1 Pitchi'** were subjected to 0.5 to 3 kR treatment and found that rooting declined from 55% to 0.5% in 3 kR. Similar trends were also observed with regard to sprouting of axillary buds, while rooting in non-treated cuttings was 85 percent.

Devaiah and Srivastava conducted experiment on mutation breeding and observed that 2.5 kR dose of gamma irradiation was close to LD-50 for *Jasminum grandiflorum* var. 'Pink Pin' and *J. flexile* species; close to 0.5 kR for var. 'Pink Thrum'; close to 1 kR for *J. calophyllum*; and 2 kR for *J. sambac*. Percentage of rooting, number of roots per cutting, length and thickness of roots decreased with increased doses of gamma radiation.

Polyploidy

a) Triploids: Spontaneous triploids with 2n = 3x= 39 chromosomes were reported by various workers in different species as given below:

Sharma & Sharma reported in *Jasminum sambac* and *J. autumnale*

Murthy & Khanna reported in *J. grandiflorum* with high usefulness and promise due to increased concrete content.

Taylor reported in *J. ilicifolium* and *J. nitidum*

Krishnaswamy (1990) reported in *J. primulinum*

b) Tetraploids: Spontaneous tetraploid with 2n=4x=52 chromosomes were reported by Dutta in J. *calophyllum*; and Raman in *Jasminum flexile*.

Varietal development

Variety	Institute	Characters
CO-1 "Pitchi" (*J. grandiflorum*)	TNAU, Coimbatore	Selection from strain of Lucknow. Buds are slightly pink shaded. Under Coimbatore conditions, it yields flower about 10 t/ha/year with 0.29 % concrete (25 kg/ha concrete).
CO-1 'Mullai' (*J. auriculatum*)	-do-	8.8 tonnes of flowers/ha with 0.34% concrete
Parimullai (*J. auriculatum*)	-do-	43 quintal/ha flowers yield and resistant to gall mite infestation
Arka Surbhi (*J. grandiflorum*)	IIHR, Bangalore	Flower yield is about 10 t/ha/year with 0.35 % concrete

Screening for disease resistance

Narayanaswamy *et al*. studied eighteen *Jasminum* species comprising of four cultivars of *J. auriculatum*; and fourteen cultivars of *J. sambac* for occurrence of anthracnose (*Colletotrichum sp*), rust disease (*Uromyces hobsoni*) and leat spot (*Cercospora jasminicola*). In another study, it was observed that *Jasminum sambac* cv. 'Eruvatchi›, *J. paniculatum* and *J. calophyllum* were completely free from diseases.

Chandra *et al.* evaluated eighteen species and eleven cultivars of *Jasminum* against yellow ring mosaic virus and conveyed that seven species *viz. J. arborescens, J. angustifolium, J. calophyllum, J. nitidum, J. suavissimum* and *J.* sambac expressed 10-100 per cent infections; whereas only five species *viz. J. beesianum*, *J. auriculatum, J. grandiflorum L.* (Pin and Thrum types), *J. auriculatum* and *J. officinalis* were found immune, indicating their highly resistance against yellow ring mosaic virus. Hence these species can be exploited for inclusion of disease resistance in breeding programme.

Another study reported that species *J. flexile, J. calophyllum, J. rigidum,* and *J. angustifolium* were tolerant to yellow ring mosaic virus at Coimbatore, but susceptible at Bangalore, indicating genotype-environment interactions. On the other hand, species *Jasminum amingay* seems to have appreciable tolerance against red spider-mite and white fly.

Chemistry of Jasmine concrete and absolute

Jasminum grandiflorum L. types are reported to contain 0.2 to 0.425 percent concrete, possesses waxes containing a series of hydrocarbons. Garnero *et al.* (1980) reported 25 hydrocarbons in the wax and the composition of the same has been reported similar to bee's wax.

Ramchandran *et al.* (1984) reported physical characters of concrete of three different species as given below.

Jasmine species	Slip point (Celsius)	Melting point (Celsius)	Refractive index (50°C)	Specific gravity (50°C)
J. grandiflorum	50	51	1.475	0.900
J. auriculatum	-	-	1.468	0.900
J. sambac	45	46	1.466	0.879

Future Research Needs

1. Effects of genotype x environment interaction needs to be studied on the quality of concrete and absolute with detailed chemical analysis in order to redefine the best cultural practices.
2. Breeding for resistance to biotic stresses with regard to prevailing diseases and insect-pests.
3. Further refinement of extraction technology for concrete and absolute to compete the quality of end product in the world market.
4. Research on oil utilization of Jasmine species other than *Jasminum grandiflorum.*

References

Goldstein (1955)

Kotyarova, M. V.1958. Dushishtykh Vashestu, 4: 156-175.

Ramchandran et al. (1984)

Sharma and Sharma (1958)

Murty and Khanna (1971).

Karmaker and Srivastava (1987).

Srivastava and Shamasundaram, 1988).

Srivastava, H. C. and Karmakar, P. G. 1986. Annual report of IIHR, Bangalore.

Srivastava and Karmakar (1988)

Srivastava, H. C. 1995. In: Advances in Horticulture. Ed. By K.L. Chadha and Rajendra Gupta. Malhotra Publishing House, New Delhi.

Suggested questions

1. What is the scientific name of Jasmine?
2. What are the other common names of Jasmine?
3. Name some other species in Jasmine.
4. Write the family to which Jasmine belongs.
5. Name a few released varieties of Jasmine.
6. What is the basic ploidy level and somatic chromosome number of major species with genomic number?
7. What breeding methodology were attempted in Jasmine.
8. What are its different uses?

17

Lemongrass (*Cymbopogon flexuosus* Nees ex. Stend Wats)

Introduction

Lemongrass grows widely in many tropical and subtropical parts of Asia, Africa and America. East Indian or true lemongrass (*C. flexuosus*) is indigenous to India and large variability occurs in the Western Ghats comprising Kerala, Karnataka and T.N., besides Sikkim and A.P. Its cultivation and distillation of oil started first in Kerala and now grown in Assam, Maharashtra and U.P. The strong lemon like odour of the oil of the grass is responsible for its nomenclature. Predominantly, there are two types of lemongrass under cultivation, *viz.*, the East Indian or Cochin or Malabar grass (*Cymbopogon flexuosus*) with superior oil quality (75-85% citral) and preferred by the industry.

The other West Indian or American lemongrass (*Cymbopogon citratus* (D.C.) Stapf. is mainly cultivated in West Indian islands, Guatemala, Brazil, Mexico and Madagaskar; and comora islands, Thailand, Bangladesh and China. It has lower citral content and lower solubility in 70% alcohol. It is indigenous to India and spread over Western Ghats, comprising Kerala, Karnataka, TN, Sikkim, Arunachal Pradesh, Assam, Maharashtra, parts of Uttar Pradesh, Punjab and Haryana.

Besides, a separate culture called Jammu lemongrass (*Cymbopogon pendulus*) is another indigenous species found wildly growing in West Bengal, Sikkim, Assam and M.P. It also has high citral content, but cultivated to a limited extent in parts of Jammu in northern India.

Origin

East Indian or Cochin or Malabar grass (*Cymbopogon flexuosus*) is considered to be native to Cambodia, Vietnam, Laos, India, Sri Lanka, Burma and Thailand, whereas West Indian or American lemongrass (*Cymbopogon citratus*) to Maritime Southeast Asia; and North Indian or Jammu lemongrass (*Cymbopon pendulus*) to Yunnan, Eastern Himalayas, Myanmar and Vietnam.

Genotypic variability

Though 54 species are reported in literature under genus *Cymbopogon*, yet the following three are commonly found in India.

Types	Scientific Name	Quality parameters
East Indian Type (Malabar or Cochin grass)	*Cymbopogon flexuosus*	Citral: 75-85% (Aldehyde) Citral-a (Geranial): 40-48% Citral-b (Neral): 30-37%
West Indian Type	*Cymbopogon citratus*	Oil content: 0.20-0.90% Citral: 60-75%
North Indian Type (Jammu lemongrass)	*Cymbopogon pendulus*	Oil content: 0.20-0.40% Citral: 75-80%

Cytogenetics

Scientific Name & Type	Chromosome No. (2n)	Reference
Cymbopogon flexuosus var. flexuosus (Red culm grass- East Indian type/ cochin/Malabar grass)	20	Lavania, 1988
Cymbopogon flexuosus var. albescens (White culm grass- East Indian type)	20	
Cymbopogon citratus (West Indian/American type)	60	
Cymbopogon pendulus (Jammu type- Jamrosa)	20, 40, 60	Babu *et. al.,* 1936

Uses

The citral, an aldehyde, extracted from the oil is an important raw material for flavouring soft drinks, in scenting soaps, detergents, as a fragrance in perfumes, cosmetics, confectionary, herbal teas, beverages and aromatherapy. Citral is a starting material for manufacture of ionones (alpha-ionone and beta-ionone). Alpha-ionone is basically used in perfumery industry, while beta-ionone is used in production of Vitamin A. In world trade, lemongrass oil is popularly known as Cochin oil, as 90 percent of the lemongrass oil is exported from Cochin port (Kerala).

Besides, citral rich oil of lemon grass has germicidal, nematicidal, medicinal and flavouring properties.

Floral Biology

Genus *Cymbopogon* belongs to family Graminae (Poaceae). It is a perennial grass with about 2 m height and profuse tillering capability. Inflorescene is a terminal panicle, highly branched into primary, secondary and tertiary ending ultimately in small paired spikes. A panicle takes 25-35 days for completion of anthesis. It is a short day plant with regular panicle emergence during October onwards in tropical conditions. It is adapted for cross pollination because of the protogynous nature of the bisexual florets. The stigma receptivity is for 3-4 hours, but maximum during first hour after opening of flower.

Nematicidal Activity

Singh *et al.*, (2009) studied nematicidal activity of oil extracted from six different genotypes of lemongrass against two root-knot nematodes (*Meloidogyne javanica and M. incognita*) and reported that genotypes HL-2, HL-3, HL-11, NLG-118, OD-19 and CKP-25 were effective for mortality against both the nematodes to varying degree i.e. 61.2 to 34.3% against *M. javanica*; and 93.6 to 74.7% against *M. incognita* at Hisar. They further reported highest mortality (71.1%) in aqueous extract of genotype CKP-25, followed by HL-11 (59.5%), NLG-118 (51.5%) against *Rotylenchulus reinformis* nematode.

Chemical composition of oil

The qualitative standards for trade of oil as specified by Bureau of Indian Standards are as below:

Character	Value
Colour & appearance	Dark yellow to light brown red mobile liquid
Odour	Lemon like
Specific gravity (30^{0}C)	0.880-0.898
Optical rotation	-3^0 to +1^0
Refractive index (30^{0}C)	1.4786 to 1.4846
Solubility in 70% alcohol	Soluble in 2 to 3 vols.
Citral content (%)	A grade – 76 to 80 %

Genetic Variability

Systematic crop improvement programme started as early as in 1951 at Lemongrass Research Station, Odakkali (Kerala) with collection of large gene pool of population samples (450 accessions), resulting in release of variety OD-19 (Sugandhi) which is still popular and under cultivation with high herb yield, high oil content and high citral percentage. Combination of these characters in one genotype appears to be due polygenic nature and heterozygosity of the genotype.

A wide range of genetic variability have been observed for different characters in lemongrass. A brief account of work done is as under:

Reference	Character	Range
Verma *et al.*, 2001	Oil content (FWB)	0.16-0.70 %
	Geraniol content	1.2-31.6 %
	Geranyl acetate	0.7-9.8 %
	Citral-a (Neral)	1.2-32.3 %
	Citral-b (Geranial)	3.7-47.3 %
	Citral content	4.9-79.6 %
Nair *et al.*, 1984	Plant height	67-124 cm
	Leaf width	0.7-2.4 cm
Kulkarni *et al.*, 1986	Tillers/plant	14-150
	Fresh herb yield/plant	91-436 g
	Oil content (FWB)	0.53-1.10 %
	Citral content	79.0-91.5 %
Pillay, 1961	Oil content (FWB)	0.08-0.29 %
	Citral content (%)	11.4-87.6 %
Kuriakose, 1995	Fresh herb yield/plant	18-364 g
	Oil content (FWB)	0.12-1.03 %
	Citral content	36.6-70.7 %
Sharma *et al.*, 1988	Oil content (FWB)	0.23-0.58 %
Yadav *et al.*, 2012	Plant height	72.9-125.7 cm
	Leaf length	53.3-89.4 cm
	Leaf width	0.7-1.5 cm
	Tillers/plant	46.7-127.0
	Fresh herb yield	241.4-701.4 q/ha
	Oil content (FWB)	0.29-1.03 %
	Oil yield (FWB)	113.5-377.4 lt./ha
	Citral content	84.1%

Yadav *et al.*, (2012) further reported that out of 12 genotypes studied, variety CKP-25 recorded the highest oil yield (377.4 lt./ha), followed by OD-58 (357.7 lt./ha), Jamrosa (348.1 lt./ha), NLG-84 (244.1 lt./ha) and HL-4 (206.7 lt./ha). On the other hand, citral content was recorded to be highest for genotype NLG-118 (84.1 %), followed by HL-2 (83.0 %), HL-3 (82.4 %), NLG-84 (81.6 %), OD-19 (81.1 %) and HL-4 (80.4 %).

Correlation and path analysis

Reference	Character	Significant correlation with		Path analysis	
		Oil content	Oil yield	Direct effects	Indirect effects
Verma *et al.*, 2003	Plant height (cm)	-ve	-ve	Oil content, fresh herb yield and leaf length	High via tillers/ plant
	Tillers/plant	+ve	+ve		
	Leaf length (cm)	-ve	-ve		
	Leaf width (cm)	-ve	-ve		
	Fresh herb yield (q/ha)	-ve	-ve		
	Oil content (%)	-	+ve		

Stability analysis

Reference	Genotype	Character	Conditions
Yadav *et al.*, 2002	HL-1 HL-2 HL-4	Fresh herb yield (q/ha)	Poor environment -do- -do-
	HL-11 Jamrosa CKP-25 NLG-84 OD-19	Oil content (%)	Good environment -do- -do- -do- Poor environment
	HL-11 OD-19 Jamrosa CKP-25 OD-58	Oil yield (lt./ha)	Good environment -do- -do- -do- Poor environment

Ideal plant type (ideotype)

The morphological characterization of 33 genotypes of lemongrass, at the Research area of Medicinal, Aromatic and Potential Crops Section (GPB), C.C.S. Haryana Agricultural University, Hisar, revealed that most of the genotypes in lemon grass were semi-spreading to compact type, tall to medium in height, with broad and green leaves having light purple leaf sheath color and medium in leaf length and tillers. About 27 genotypes were non-flowering and only six genotypes (NLG-1, Chirharit, RRL-16, Karishna, CKP-25 and OD-58) were able to bear flowering. On the basis of one or more morphological characters, each genotype is different from others (Arya *et al.*,2024).

1. Erect growing habit of the plant with quick regeneration capacity.
2. Long and broad leaves, as these contain more oil grands in comparison to short and narrow leaves.

3. High tillering capacity of the plant/clump, as it is directly related to increased herb yield.
4. Non-hairy leaves are preferred over hairy leaves due to easy harvest.
5. High oil content during dry season (>0.60%) with high citral content (>80%).

Compact — Semi-compact — Semi-spreading — Spreading

Lemongrass ideotypes

Breeding Strategies

The varietal performance largely depends upon its genetic potential besides its interaction with the prevailing environments. Therefore, genetic amelioration keeping in view the need of the cultivators under specific situations and utilizing industry for certain specific trait. Hence, breeders have to tailor specific varieties for specific situations.

Keeping in view, the allogamous nature (cross pollination) and vegetative propagation of lemongrass, the following breeding approaches can be applied for improvement of high oil yield and oil quality.

Genetic collection and evaluation

Collection of seeds/planting materials from its habitat and different areas of cultivation make the plant genetic resource (PGR) a strong base for studies on variability, heritability, correlation, genetic advance and ideal plant type (ideotype). For any crop improvement programme, a wider genetic base is a pre-requisite. Therefore, widening of genetic base through introduction and collections can be instrumented in bringing new vistas in the production of high oil yielding and oil quality varieties, besides resistance to diseases and insect-pests. There is no use of exploration, if the collection is preserved in perpetuity and not properly evaluated and utilized in the crop improvement programme.

Clonal selection

Lemongrass is propagated by clones/vegetatively. Therefore, clonal selection can be beneficially utilized by vegetative propagation, as there is no further segregation. Hence desired/selected clone can be effectively used for a particular or combination of characters. Similarly, Pragati from OD-19 from clonal selection.

Chirharit from Cauvery; and Krishna from SD-68 were produced from open pollinated seed progenies in lemongrass.

Hybridization (intra-and inter-specific)

Keeping in view the viable and profuse seed setting, the technique of hybridization among improved and diversified clones/types is important to generate diversity. Similarly, wide crosses are also desirable to transfer resistance against biotic and abiotic stresses from related species. The inbred lines developed by continuous selfing can be crossed among themselves after evaluating for general combining ability (GCA) and the heterosis (hybrid vigour) thus obtained can be exploited by taking advantage of vegetative propagation in these crops.

Mutation

Development of useful mutants through gamma irradiation and chemical mutagens have been reported by Shylaraj *et al*., 1988 in *C. flexuosus*; and Kak *et al.,* 1993 in Jamrosa (*C. pendulus*).

Polyploidy

Keeping in view of the somatic chromosomes from diploid to hexaploid in the genus *Cymbopogon*, and importance of vegetative parts, efforts should be made to crate polyploidy in crop by using colchicine and by other means to utilize variants thus created by doubling the chromosome number or by combining chromosome sets from different species in their amphidiploids.

Population improvement methods

Recurrent selection is very useful method for population improvement and it is practiced in a synthetic variety or composite population to improve yield and quality by concentration of additive variances.

Stability analysis

The stable genotypes of lemongrass for different characters over different spacing environments (60 x 60, 60 x 45 and 45 x 45) at Research area of MAP Section (GPB), CCS HAU, Hisar revealed the presence of both linear and non - linear G x E interactions. The results on oil content % (FWB) revealed that, out of all the thirty-three genotypes / varieties, only twelve genotypes exhibited stable performance with high mean. NLG-4, Krishna, NLG-5, NLG-118, OD-58, NLG-84, HL-11, RRL-16 and CKP-25 were found best genotypes for oil content % (FWB) viewing high mean performance with above average (bi >1) responsiveness; and genotype, OD-19, OD-23, and OD-388 were suitable for favourable environments, none were suitable for poor environments (Singh *et al.*, 2024).

Biotechnological approaches

Somaclonal variation: Somaclonal variants generated through tissue culture could also lead to the development of a new cultivar.

Varietal Achievements

The following crop varieties of different aromatic plants have been developed from various institutes utilizing different breeding approaches.

Variety	Name of Institute	Oil quality	Methodology/Species
Sugandhi (OD-19)	KAU, Odakkali	Citral (85-88%)	
OD-440 (Red stem)	KAU, Odakkali	Citral	
OD-408 (White stem)	KAU, Odakkali	Citral	
RRL-82	RRL, Jammu	Citral	*C. pendulus*
CKP-25	RRL, Jammu	Citral (82-85%)	*C. khasianus* x *C. pendulus*
RRL-16	RRL, Jammu	Citral (80%), Oil (0.6-0.8% FWB)	*C. pendulus*
RRL-14	RRL, Jammu	Citral (70%)	*C. khasianus*
Praman (Clone 29)	CIMAP, Lucknow	Citral (82%)	C. *pendulus* (Clonal selection)

Pragati (LS-48)	CIMAP, Lucknow	Citral (85%), Oil (0.63% FWB))	Clonal selection from OD-19
Prasoon	CIMAP, Lucknow	Citral (89%)	-
Krishna	CIMAP, Banglore	Citral (81%)	OPS* from SD-68
Cauvery	CIMAP, Bangalore	Citral (84%)	-
Chirharit	CIMAP, Pantnagar	Citral	OPS* from Cauvery
NLG-84	NDUAT, Faizabad	Citral	-
CIM-Bio 171	CIMAP, Lucknow	Geraniol (21%)	-
Nima	CIMAP, Lucknow	Citral (85-90%)	-
CIM-Shikhar	CIMAP, Lucknow	Citral (80%)	-
Jam Rosa	RRL, Jammu	Oil (0.3-0.4%)	Interspecific hybridization with *C. jawarancausa*
GRL-1		Rich in geraniol	-
RRL-39	RRL, Jammu		-

*OPS-Open pollinated seeds

References

Arya R.K., V. Kumar, P.K. Verma and P. Kumar. (2024). Indian J. Plant Genet. Resour. 37(3): 433-438.

Babu, C.N. 1936. Curr. Sci., 4: 739.

Kulkarni, R.N., K. Rajagopal and S. Ramesh. 1986. Plant Breeding, 97: 78-81.

Lavania, U.C. 1988. Cytologia, 53: 517-524.

Nair, R.V., N. Singh, V. Singh and K.S. Anathn. 1984. Indian Perfumer, 28: 129-132.

Piliay, K.P. 1961. Farm Bulletin No. 16, Ministry of Food and Agriculture, New Delhi, p. 26.

Sharma, J.R., A.A. Naqui, R.K. Lal and H.O. Mishra. 1988. Genetica Agraria, 42(1): 13-23.

Shylaraj, K.S. and J. Thomas. 1988. Indian Perf, 32(2): 116-119.

Singh R, P.K. Verma, R.K. Behl and R.K. Arya (2024). Ekin J. 10(1): 22-26.

Singh, Y.K., V.K. Madan and Sewak Ram. 2009. Abstracted in Proceedings of Workshop on "Prospects and Problems in Production, Processing, and Marketing of Medicinal and Aromatic Plants in Haryana" organized at MA&UUP Section, Dept. of Plant Breeding, CCS HAU, Hisar on March 20-21, 2009, p. 86 & 87.

Verma, P.K., M.K. Deen, S.N. Gupta and C.S. Tyagi. 2001. Indian J. Plant Genet. Resources, 14: 173-174.

Verma, P.K., M.L. Saini, O.P. Yadav, C.S. Tyagi and V.K. Madan. 2003. In: Proceedings of 1st National Interactive Meet on Medicinal and Aromatic Plants, CIMAP, Lucknow, pp. 114-118.

Verma, P.K., Ravinder Singh, S.N. Gupta and C.S. Tyagi. 2005. In: Course Compendium of Winter School on "Advances in Medicinal, Aromatic and Under-utilized Plants Research" held at CCSHAU, Hisar from Sept. 29 - Oct.19, pp. 48-53.

Yadav, O.P., P.K. Verma, V.K. Madan and C.S. Tyagi. 2002. In: Symposium on "Perspective in Medicinal and Aromatic plants- Research and Development" held at CCS HAU, Hisar on March 7-8, 2002, pp. 33-36.

Yadav, O.P., I.S. Yadav, V.K. Madan, P.K. Verma and M.K. Deen. 2012. Abstracted in Proceedings of Seminar on "New perspectives in Aromatic and Medicinal Plants" held at CCS HAU, Hisar on Feb. 8-9, 2012, p.145-146.

Suggested questions

1. What is the common scientific name of Lemon grass?
2. What are the other species of Lemon grass?
3. To which family Lemon grass belongs.
4. Name a few released varieties of Lemon grass.
5. What is the ploidy level and somatic chromosome number of major species?
6. What are the effective breeding methodology and why?
7. What are different uses of its essential oil?

18

Mint/Pudina (*Mentha* Species)

Introduction

Japanese mint (*Mentha arvensis*) was first introduced in India at Regional Research Laboratory, Jammu during 1952 from Japan (Kapoor *et al.*, 1955); and then to Drug Research Laboratory, Jammu at Chakrohi farm. Subsequently, it was introduced to CIMAP Northern Zonal Research Station at Haldwani in Nainital district (U.P.) during 1963-64. In India, the crop is mainly grown in Tarai districts of U.P. (Nianital, Badayun, Bilaspur, Rampur, Moradabad and Bareilly); extending to parts of Indo-Gangetic plains (Barabanki and Lucknow); and parts of Punjab (Ludhiana and Jalandhar) in the north-western India. Mint or Pudina comes under a small genus *Mentha* of aromatic perennial herbs which belongs to family Labiatae (Laminaceae), distributed mainly to temperate and sub-temperate regions of the world. It has blue, laverder or white flowers in whorls.

Peppermint (*M. x piperita*) and spearmint (*M. spicata*) are largely cultivated in America, mainly in Oregan, Indiana, Idaho and south Michigan states. It was introduced in India during early 1980s.

Uses

The above-ground herb (foliage) on distillation yields essential oil containing a large variety of aroma chemical command a large and world-wide demand in industry as given below.

The Japanese mint i.e. *M. arvensis* is the principal source of menthol in the world due to presence of high percentage of menthol content over peppermint (*M. x piperita*). It has, therefore, a bitter cooling taste and harsh odour. It is used in cold remedies prescriptions *viz.*, cough drops, dentrifices, mounth washes, cosmetics, scenting cigarettes, flavouring tabacco, chewing pan etc.

The peppermint oil obtained from *M. x piperita*, on the hand has a fine odour and is an excellent carminative, antiseptic, preservative and gastro-stimulant properties. As a flavouring agent, it is used in a wide range of pharmaceuticals, confectionery, alcoholic drinks, dental creams and mouth washes.

The oil of spearmint (*M. spicata*) is rich in carvone, a digestive and gastro-stimulant compound. It has large demand in confectionery, chewing gum and tooth paste industries.

The oil of bergamot mint (*M. citrata*) is rich in linalyl acetate and linalool; and thus used in cosmetics industry.

Genotypic variability

In India, six species are reported to grow wild in hilly parts of north-western Himalayas and attempts have been made to examine the wild growing populations for their economic utilization in essential oil industry.

These species are *Mentha arvensis* (Japanese mint), *M. piperita* (Peppermint), *M. citrata* (Lemon/bergamot mint), *M. spicata* (Spearmint), *M. cardiac* and *M. viridis* etc. But the former four species are well known and under cultivation on commercial scale.

Besides, *M. longifolia* (Horse mint), *M. sylvestris*, *M. aquatica* (Water/Marsh mint), *M. vercillata* (Tea mint), *M. suaveolens* (Apple mint) and *M. pulegium* L. (Pole mint) have also been reported in the literature. However, a number of taxonomical valid species in the genus is a matter of speculation, as these crossed freely amongst themselves, producing innumerable intermediary forms; and some of these have polyploidy also playing an important role in speculation.

1. ***Mentha arvensis*** var. piperascense (Japanese mint): The most popular as it is the principal source of menthol, methyl acetate and menthone, used in prescriptions for cold remedies, cough drops, dentrifices, mouth washes, cosmetics, scenting cigarettes, flavouring tobacco, chewing pan etc.
2. ***Mentha piperita*** (Peppermint): The oil has fine odour and is an excellent carminative, antiseptic, preservative and gastro stimulant properties.
3. The other species *viz.*, ***Mentha spicata*** (Spearmint) and ***Mentha citrata*** (Bergamot or lemon mint) are also used in wide range of pharmaceutical, confectionary, alcoholic drinks, cosmetics, dental creams and mouth washes.

Common Name	Scientific Name	Quality parameters
Japanese mint	*Mentha arvensis*	Menthol: 80-95% L- menthone: 6.8-22.2% Methyl acetate: 14.1-24.4%
Spearmint	*Mentha spicata*	L- carvone: 55-60%
Peppermint	*Mentha piperita*	Menthol
Bergamot/ lemon mint	*Mentha citrata*	Linalool

Mentha spicata

Mentha arvensis

Mentha piperita

Mentha citrata

Cytogenetics

Sobti (1962) reported that among 49 recognizable Texas of genus *Mentha*, belonging to family Labiatae have chromosome numbers varying from 2n=20 to 120 with two basic number of x=5 and x=6.

Sobti (1969) further studied inter-relation between 16 species, 33 varieties and 100 geographical races assembled at Jammu and concluded that morphological characters and polyploidy have evolved independently with no correlations between quality of oil and appearance of the foliage in a geographical race, but relationship exists between chromosome constitution and oil content. He further reported parentage and cytogenetics along with quality composition of different cultivated texas of *Mentha* as under:

Species	Pedigree	Chromosome No. (2n)	Characteristics
Spearmint (*M. spicata*)	*M. longifolia* x *M. rotundifolia*	48	High carvone content (59-68%)
Scotch mint (*M. gentilis*)	*M. spicata* x *M. arvensis*	-	High carvone (53.9-68.0%) & rust resistance disease.
Peppermint (*M. piperita*)	*M. spicata* x *M. aquatica*	-	Two commercial varieties under cultivation.
Black Mitcham	*M. piperita* var. vulgaris	76	Vigorous growing, hardy & high oil content.
White mint	*M. piperita* var. *officinalis*	36	Less hardy with green stem and leaves, poor yielder but high oil quality; poor herb yield.
Japanese mint (*M. arvensis)*	*M. arvensis* x *M. aquatica*	96	Main source of Menthol with a bitter cooling taste and odour.
Bergamot mint (*M. citrata)*	*M. aquatica* x *M. viridis*	96	Robust in growth like Japanese mint with high linalool and linalyl acetate.
Pennyroyal	*M. pulegium*	20	High pulegone and used as precursor of menthol.
Pudina (*M. spicata* var. *crispa*)	Syn. to *M. viridis* L.	-	Round to broad ovate leaves, grown in kitchen garden for chutney making, low in oil content but with high carvone content.

Crop Improvement

Due to small size of chromosomes of *Mentha* species, it is helpful in widening its capacity for adaptability and its world-wide distribution. The volatile oil in mint crops occurs in numerous hairy gland on the upper and under surface of the leaves but the stem part has generally negligible quantity. The oil content increase rapidly during high vegetative and physiological active growing flowering stage during the period of May-June at first harvest; and second harvest in September-October. The oil content rapidly declines after full bloom stage.

An ideal plant type of *Mentha* should have the following features

- An erect plant type with profuse branching nature.
- Producing vigorous bushy growth with higher leaf: stem ratio.
- The number of glandular hair per unit area on the leaves should be more, as it provides an index of high oil content.
- It should produce high oil content with high (free) menthol content.

Germplasm collection

CSIR-CIMAP, Lucknow has collection of six *Mentha* species along-with 55 mutants, 369 strains and a large number of cultivars developed so far in their field gene bank. Besides, several collections are being maintained at CSIR-RRL, Jammu; Department of Forest Product Utilization, YSP Univ. of Horticulture & Forestry, Solan (H.P.); and Department of Agronomy, PAU, Ludhiana.

Inter-specific hybridization and clonal selection

In India, the first variety, named, EC-41911 released by NBPGR, New Delhi under the species *M. arvensis*, was selection from progeny of segregating population of inter-specific cross (*M. arvensis* var. *piperascens x M. piperita*) received from USSR, having high herbage yield, high oil content and high oil quality (high menthol content-70%).

Sahu (1993) obtained two clones rich in carvone from an inter-specific cross between *M. aevensis* var. piperascens x *M. spicata*.

Euploidy

Tucker and Fairbrothers (1981) reported an euploid series through inter-specific cross between *M. arvensis* (2n=72) x *M. spicata* (2n=48).

Mutation breeding

- Murray (1971) incorporated wilt resistance through crossing developing **Murray Mitchum** through mutation breeding.
- A few mutants have been obtained and released as a cultivar through physical and chemical mutagens in India and abroad.
- Kar and Kaul (1976) reported LD-50 as 4.0 kR for *M. arvensis*; 3.0 kR for *M. citrata*; and 2.0 kR for *M. piperita*.

Polyploidy

Ammal (1962) was the first India to evolve a **tetraploid** with 2n=192 chromosomes in Japanease mint (*M. arvensis*) through cholchiploidy and named it as **Jammu mint**. It was reported to have 5.0 percent oil content on dry weight basis, as against 3.0 percent of the normal diploid parent. But it could not come to commercial level due to inconsistency of yields.

Bradu *et al.* (1971) later reported that it was actually due to the fact of having low menthol content and more of menthone content. It was also found susceptible to fungal and viral diseases.

Genetic variability, heritability and correlation

Heritability estimates for different characters among 15 genotypes of *Mentha arvensis* were reported as below:

Character	Heritability (%)	Reference
Leaf/stem ratio	61-69	Sharma and Tyagi, 1990
Menthol content (%)	93-96	
Oil content (%)	94-98	

Sharma and Tyagi, 1991 studied heritability, genetic advance and correlation with oil yield.

Character	Heritability	Genetic advance	Characters	Correlation with oil yield
Leaf/stem ratio	Lowest	Lowest	Oil content vs oil yield	+ive
Oil content (%)	High	Very low	Menthol vs oil yield	+ve
Menthone content (%)	High	Very low	Menthone vs oil yield	-ve
Menthol content (%)	High	Highest	Menthone vs menthhol	-ve
Herb yield (kg/ha)	-	High		

Kumar *et al.*, 2004 studied the genetic variation in open pollinated seed progeny (OPSP) of *Mentha piperita* cv. Kukrail and reported maximum variation for menthol and menthofuron content.

Character	OPSPs (ff homozygote)	cv. Kukrail (Ff heterozygote)
Menthol content (%)	30-77%	30.0-38.3%
Menthofuron content (%)	13-35%	5.6-8.5%

Verma *et al.*, 2003 reported prevalence of sufficient genetic variation and correlation and path-coefficient among different genotypes in four *Mentha* species as under:

Character	GCV (%)	Heritability (%)	Genetic gain (%)
Plant height (cm)	23.58	87.99	45.57
Fresh herb yield (kg/ha)	57.16	97.09	114.67
Oil content (%)	27.85	97.54	56.03
Oil yield (kg/ha)	67.24	96.97	136.19

Correlation and Path analysis

Character	Direct effects	Indirect effects	Correlation with oil yield
Plant height (cm)	0.034	Fresh herb yield had highest indirect effects via plant height and oil content.	0.71
Fresh herb yield (kg/ha)	0.772		0.94
Oil content (%)	0.034		0.66

Variability among *M. arvensis* genotypes/varieties at Hisar

Genotype	Fresh herb yield (q/ha)*	Oil content (%, FWB)	Oil yield (lt./ha)	Menthol content (%)
Gomti	81.7	0.92	75.16	
Shivalik	80.0	1.20	96.0	72.30
Himalayan	121.7	1.10	133.87	75.60
EC-41911	76.7	1.10	84.37	
HMA-1	140.0	0.94	131.60	

*Performance based on two cuttings under limited water conditions. (Yadav *et al.*, 2012.)

Chemical constituents of different *Mentha* species

Species	Chemical constituents
M. piperita	Menthol (48-60%), Menthone (9-12%), Limonene (2-5%), 1,8-cineole (7.3%), Methyl acetate (3.6%)
M. spicata	Carvone (50-65%), Limonene (2.5%), Myrecene (2.5%), 1-8 cineole (2.6%), Methyl acetate (2.0%)
M. citrata (Syn. M. aquatica)	Linalool (30-50%), Linalyl acetate (35-45%)
M. arvensis	Menthol (65-85%), Menthone (7-20%), Methyl acetate (7-20%)

Breeding Methodology

Perusal of literature on the *Mentha* species showed that in these cross pollinated species a number of classical breeding approaches, namely, clonal selection, interspecific hybridization, induced mutagenesis and polyploidy breeding have been employed to genetically improve. As *Mentha* is commercially propagated through vegetative means, their intentional crossing can be readily fixed and made usable through vegetative mean without going for homozygotization. Therefore, more powerful hybridization approaches like gene introgression, somatic hybridization and double haploid seems to be significant in the improvement of species.

Breeding methodology used and varietal achievements

Variety	Breeding Methodology
Mentha piperita	
Kukrail	Mutation breeding- Wider adaptability, low disease intensity
Pranjal	Clonal selection-
Tushar	Mutation breeding- Partial shading conditions
CIM-Indus	Selected from open pollinated seed raised progeny of variety Kukrail
CIM-Madhuras	Selected from open pollinated seed raised progeny of variety Kukrail

***Mentha arvensis* (Japanese mint)**	
MAS-1	Selection from Thai bud sprout (1883)
Hybrid-77	Vigorous culture evolved through cross between MAS-1 x MAS-2
EC-41911	Progeny selection from inter-specific cross *M. arvensis* x *M. piperita*
Shiwalik	Introduction from China
Saksham	Tissue culture selection from Himalayas
CIM-Saryu	-
CIM-Kranti	Cold & frost toerant, developed through half sib progeny selectioncan be grown during winter
CIM-Vashisht	Used in Aromatherapy
Kushal	Developed through tissue culture, free from diseases & pests
***Mentha spicata* (Spearmint)**	
MSS-1	Selection from germplasm introduced from USA
MSS-5	Clonal selection from MSS-1
Punjab Spearmint-1	Clonal selection
Neerkalka	Inter-specific hybrid (*M. arvensis* cv. Kalka *x M. spicata* cv. Neera)
***Mentha citrata* (Bergamot mint)**	
Kiran	Mutation breeding (gamma irradiation)-1988

Varietal achievements and oil quality: The following varieties have been released of different species of *Mentha*.

Mentha species	Variety	Institute	Oil content & Quality
M. arvensis (Japanese mint)	MAS-1 Hybrid-77 EC-41911 Shiwalik Saksham Sambhav Damroo Kosi CIM-Saryu CIM-Kranti CIM-Vashisht Kushal	CIMAP CIMAP ICAR CIMAP CIMAP CIMAP CIMAP CIMAP CIMAP CIMAP CIMAP - -	Oil (0.8-1.0%), Menthol (81%) Menthol (81.5%) Menthol (70%) - Menthol (85-90%) - - Menthol (75-78%) Menthol (78-80%) Menthol (80%) Pulegone (65-68%) -
M. piperita (Peppermint)	Kukrail Pranjal Tushar Pratik CIM-Indus CIM-Madhuras	CIMAP CIMAP CIMAP CIMAP CIMAP CIMAP	Oil (0.40%) - Menthol (33%), Menthone (27%) - Menthofurone (27%) -

M. spicata (Spearmint)	MSS-1 MSS-5 Punjab Spearmint-1 Ganga Neerkalka	ICAR ICAR ICAR CIMAP CIMAP	Carvone (60%) Carvone (70%) Oil (0.57%), Carvone (68%) - Carvone (64-76%), Menthol (0.66-2.45%)
M. citrata (Bergamot/ Lemon mint)	Kiran	CIMAP	Linalool (48%)

Future research needs

- Development of variety as per environment to have high oil quality.
- Resistance breeding against common biotic stress (diseases and pests).
- Resistance breeding against abiotic stress (low irrigation & high pH).

References

Ammal, Janki. 1962. Current Sci., 3(10): 387-388.

Bradu, B.L., S.G. Agarwal, V.N. Vasista and C.K. Atal. 1971. Planta Medica, 20(3): 219-222.

Anonymous. 2002. In: Aromatic and Medicinal Plants (Shiva et al., Eds.) International Book Distributors, Dehradun, India.

Gorden, P. De. Wolf.1954. Baileya, 2 (1): 3-10.

Kak, S.N. and B.L. Kaul.1981. Proc. Indian Acad. Sci., 90 (3): 211-215.

Khanuja, S.P.S., Sushil Kumar, A.K. Shasany, Sunita Dhawan, M.P. Darokar, A.A. Naqui; O. P. Singh and N.K. Patra. 2001. J. of Med. and Arom. Plants Sciences, 23 (2): 110-112.

Kumar, B., S.P.S. Khanuja and N.K. Patha. 2004. J. of Med. and Arom. Plants Sciences, 26 (1): 84-88.

Patra, N.K., M. Tanweer, S.P.S. Khanuja, A.K. Shasany, M.P. Singh, V.P. Singh and S. Kumar. 2001. Theoretical and Applied Genetics, 102 (4): 471-476.

Sharma, S. and B.R. Tyagi. 1990. J. of Genetics and Plant Breeding, 44 (2): 81- 83.

Sharma, S. and B.R. Tyagi. 1991. J. of Genetics and Plant Breeding, 45 (4): 257-262.

Sobti, S.N. 1962. Bull. Reg. Res. Lab., Jammu, 1 (1): 40-41.

Sobti, S.N. 1969. CSIR News, 19 (10): 3-4.

Verma, P.K; O.P. Yadav; C.S. Tyagi; G. Talwar and V. K. Madan. 2003. In: Proceedings of National Seminar on "Emerging Trends in Research and Business Management of M&APs" held at CCS HAU, Hisar from March 11-13, pp.26-29.

Verma, P.K., Ravinder Singh, S.N. Gupta, O.P. Yadav and C.S. Tyagi. 2005. In: Course Compendium of Winter School on "Advances in Medicinal, Aromatic and Under-utilized Plants Research" held at CCSHAU, Hisar from Sept. 29 - Oct.19, pp. 55-62.

Yadav, O.P., I.S. Yadav, V.K. Madan, P.K. Verma and M.K. Deen. 2012. Abstracted in Proceedings of Seminar on "New perspectives in Aromatic and Medicinal Plants" held at CCS HAU, Hisar on Feb. 8-9, 2012, p.145.

Suggested questions

1. What are the common species of mint/pudida?
2. To which family mint/pudida belongs.
3. Name a few released varieties of mint/pudida.

4. What is the ploidy level and somatic chromosome number of Spearmint and Japanese mint?
5. What are the effective breeding methodology and why?
6. What are different uses of its essential oil?

19

Palmarosa [*Cymbopogon martinii* (Roxb.) Wats. Var. motia Burk.]

Introduction

Palmarosa/Roshagrass (*Cymbopogon martinii* var. motia): It is a tall perennial tuffed hedge grown naturally as well as commercially in the old world tropics and sub-tropics of India. It occurs in open scrub forests in parts of MP, Maharashtra and AP, where it is commercially collected and distilled for its oil. It is also sporadically found in Karnataka, TN and part of UP. The herb along with flowering tops and foliage, on distillation yield a colourless to pale yellow essential oil with sweet scented rose-like aroma, containing high geraniol; and used in cosmetics, toiletry and perfumery industry. Concerted efforts during last two decades to develop better cultural practices and development of improved varieties, the area has expanded all over the central, western and southern states of India. This crop is commercially grown in Indonesia, East African countries, Cuba and Brazil. At present, the total oil production is estimated to be around 300 tonnes annually, and two-third (200t) of it comes from cultivation.

Ginger grass (*Cymbopogon martinii* Stapf. var. sofia): This is another form of roshagrass. Its distribution is overlapping in forests with the ture palmarosa grass, although it has preference for more moist-substratum. Both the forms are morphologically identical in field except their odour when leaves are crushed. Ginger grass has dense tuft and more luxuriant growth, but produces shorter inflorescence and more tolerant to the stagnating water. It crosses freely with palmarosa in nature, producing many intermediary forms. Therefore, pure line seed by selfing/bagging of the inflorescence before anthesis is suggested for cultivation, and plants at flag leaf stage are rouged to avoid any mixture of ginger grass in palmarosa field to maintenance the high quality of palmarosa oil with a pleasant sweet rosaceous green note.

Scientific Name	Oil odour	Quality parameters
Cymbopogon martinii var. motia	Rose like	Geraniol: 75-90% Geranyl acetate: 5.5% Linalool: 2.54%
Cymbopogon martinii var. sofia	No fine smell	Geraniol: 39-48% Dipentine: 24.1% Carvone: 20% Dihydro carvone: 10% Carvecol: 13.5% Perillyl alcohol: 14%

Floral biology

The genus *Cymbopogon* is unique and majority of the species are aromatic grasses with complex inflorescence and mainly distributed in the tropical south-east Asia and Africa. Out of total 102 species, 56 are found in Africa and 21 in India. Palmarosa is highly heterogenous due to open pollination, where 25-35 percent florets in the inflorescence are hermaphrodite (bisexual) and the rest are staminate (male). Pollen fertility in the florets vary from 84-96 percent.

Low seed setting in the selfed inflorescence may be due to:

i. Presence of self- incompatibility system

ii. Due to some extent of protogyny

iii. Period of pollen viability and duration of stigma r

Genotypic and Cytogenetic variability

Crop	Scientific Name	Chromosome No. (2n)	Reference
Roshagrass/ Palmarosa	*Cymbopogon martinii* var. motia	20 (Diploid)	Rao and Sobti, 1987
Ginger grass	*Cymbopogon martinii* var. sofia	40 (Tetraploid)	

Genetic variability, Heritability, Correlation and Path Analysis

A wide range of genetic variability, heritability (broad sense) and correlations have been observed for different characters based on 17 genotypes in palmarosa. A brief account of work done is as under (Punia et al., 1986):

Character	Range	Heritability % (broad sense)	Correlation with oil yield
Plant height (cm)	163-244	48.4	0.57*
Tillers/plant	22.8-57.8	78.6	0.56*
Leaf length (cm)	25.5-36.7	39.0	0.87**
Leaf breadth (cm)	1.72-2.58	28.2	0.61**
Inflorescence length (cm)	52.8-76.8	40.1	0.42*
Leaf weight/clump (g)	56.7-167.5	40.2	0.88**
Stem weight/clump (g)	183.3-570.0	35.0	0.92**
Inflorescence weight/ clump (g)	81.7-236.7	53.6	0.87**
Leaf+Inflorescence/ stem ratio	0.53-0.90	63.8	-0.45*
Oil content (%) FWB	0.28-0.41	66.3	0.33
Fresh herb yield/ clump (g)	323.3-952.5	37.1	0.96**
Oil yield/clump (ml)	1.08-3.61	44.0	-

Path analysis

A study based on 31 genotypes of palmarosa revealed high direct effects of fresh herb yield/plant followed by stem weight/clump and oil content besides high indirect effects except plant height. Highly significantly genotypic correlations were also observed by all the traits with oil yield/clump as given below (Verma *et al.*, 1990):

Characters	Direct effects	Indirect effects	Genotypic correlation with oil yield/clump
Tillers/clump	0.011	Highindirect effects of fresh herb yield/ clumpvia all characters except plant height	0.68**
Inflorescence length (cm)	0.017		0.56**
Inflorescence weight/ clump (g)	0.085		0.87**
Leaf weight/clump (g)	0.125		0.76**
Stem weight/clump (g)	0.302		0.97**
Fresh herb yield/ clump (g)	1.373		0.97**
Oil content (%) FWB	0.230		0.49**

Yadav *et al.*, 2012 studied 64 clonal selections of palmarosa on single plant basis for various morphological traits at Hisar under AICRP on M&APs. The results revealed that plant height ranged from 140 cm (RH-03-6) to 250 cm (RH-03-38); number of internodes from 7 (RH-03-6, 14, 22 and 29) to 11 (RH-03-8, 43, 46, 47, 48); internode length from 13 cm (RH-03-37) to 28 cm (RH-03-9, 57); flag leaf length from 6.5 cm (RH-03-1) to 23 cm (RH-03-26, 64); flag leaf width from 0.5- 2.0 cm; leaf blade length from 15-36 cm; leaf blade width from 1.3-3.0 cm; tillers/plant from 4.0-66 (RH-03-66); panicle length from 32 to 98 cm (RH-03-52); panicles/plant from 3 to 19; and oil content on fresh plant basis from 0.20 (RH-03-28) to 0.74 (RH-03-20). The computed oil yield /plant (FWB) ranged from 9.52 (RH-03-46) to 98.9 ml (RH-03-43). It was followed by RH-03-29 (81.8 ml), RH-03-64 (80.5 ml), RH-03-22 (76.2 ml), RH-03-15 (74.4 ml) and RH-03-11 (67.2 ml). For oil quality, geraniol content varied from 9.90 to 82.50 %, where it was the highest for RH-03-11 (82.50 %), followed by RH-03-62 (80.80 %), RH-03-24 (80.40 %), RH-03-31 (79.60 %) and RH-03-17 (79.00 %).

Breeding objectives

1. High biomass yield per unit area.
2. High oil content (0.53 %) and oil quality (> 85% geraniol content).
3. Resistant to frost (during winter season) and biotic (diseases and insect-pests) and abiotic stresses (tolerance to water stagnation/submergence during monsoon season and drought/water stress conditions during summer).

Ideal plant type (ideotype)

Character	Magnitude	Remarks
Plant height	200-250 (cm)	The plant should be tall with long and compact inflorescence; Long and broad leaves; and high L+I/stem weight ratio (>0.6)
Tillers/plant	40-60	
Leaf length	25-35 (cm)	
Leaf width	2.4-2.6 (cm)	
Inflorescence length	50.75 (cm)	
L+I/stem weight ratio	0.6-0.7	
Oil content (FWB)	> 0.35%	

Punia *et al.,* 1986

Approximate oil content and oil quality in different plant parts

Part	Oil content (FWB-%)	Geraniol content (%)	Geranyl acetate (%)
Whole plant	0.53	72.4-86.5	6.3-15.4
Inflorescence	0.83	75.5-84.2	10.6-24.8
Leaves	0.58	92.9-94.6	4.5-5.0
Stem (DWB)	0.03	-	-

FWB- Fresh weight basis; DWB- Dry weight basis

Pareek *et al.,* (1981a) reported higher oil content in inflorescence; good oil quality in leaves; and negligible oil content in stem, both the plant parts (leaves and inflorescence) should be in higher proportion in comparison to stem i.e. leaf + inflorescence weight/ stem weight should be higher as above.

Therefore, an ideal plant type must fulfil the following conditions

1. Erect growing habit of the plant with quick regeneration capacity.
2. Long and broad leaves contain more oil glands.
3. High tillering capacity of the plant, as it is directly related to increased herb yield.
4. Non-hairy leaves are preferred over hairy leaves due to easy harvest.
5. High oil content during dry season (>0.60%) with high geraniol content (>80%).

Stage of harvest

Pareek *et al*., (1981b) reported that the most suitable stage of harvest is early seed formation stage (milk/dough), when majority of inflorescence flowers colour changes from green to brown for obtaining good quality oil.

Stage of harvest	Herbyield (q/ha)	Oil content (%, FWB)	Oil yield (l/ha)	Odour
Most flower open	314	0.49	157	Terpenic rosaceous
Early seed formation	272	0.50	136	Rosaceous green
Late seed formation	240	0.44	105	Rosaceous

Breeding Strategies

The varietal performance largely depends upon its genetic potential besides its interaction with the prevailing environments. Therefore, genetic amelioration keeping in view the need of the cultivators under specific situations and utilizing industry for certain specific trait. Hence, breeders have to tailor specific varieties for specific situations.

Keeping in view the allogamous nature (cross pollination) and vegetative propagation of palmarosa, the following breeding approaches can be applied for improvement of high oil yield and oil quality.

i. **Genetic collection and evaluation:** Collection of seeds/planting materials from its habitat and different areas of cultivation make the plant genetic resource (PGR) a strong base for studies on variability, heritability, correlation, genetic advance and ideal plant type (ideotype). For any crop improvement programme, a wider genetic base is a pre-requisite. Therefore, widening of genetic base through introduction and collections can be instrumented in bringing new vistas in the production of high oil yielding and oil quality varieties, besides resistance to diseases and insect-pests. There is no use of exploration if the collection is preserved in perpetuity and not properly evaluated and utilized in the crop improvement programme.

A systematic collection and its evaluation was made earlier by several workers and shown sizable range of variability in yield and its component traits (Gupta *et al.*, 1980; Sahoo and Dutta, 1982 and Sethi *et al.*, 1986). As a result, IW-31243 and IW-31245 were identified as superior selections with higher herb and oil yield per cutting. Later on, after evaluation in MLT over years, IW-31245 was found better and released for general cultivation.

ii. **Clonal selection:** Palmarosa can be propagated both vegetative and by seeds. The viable seeds thus produced under varying environments are highly heterozygous. Therefore, clonal selection can be beneficially utilized by vegetative propagation, as there is no further segregation, if desirable clone is selected for a particular or combination of characters. In palmarosa, RH-49, RH-17 genotypes were initially selected and finally RH-49 was released in 1989 by CCS HAU, Hisar.

Similarly, RRL(B)-49, 65, 69, 71 and 77 were developed by RRL, Bhubaneshwar. Bradu and Atal (1974) also made a selection from the progeny of *C. khasianus* to identify geraniol rich line (RRL-14) with 70 percent geraniol.

Sobti *et al.*, 1978 made selections from the progeny of *C. nardus* var. confertiflorus for high geraniol and geranyl acetate rich single line.

Sharma *et al.*, 1987 initially developed four inbreds and then developed elite bulk from polycross nursery, which was later on released as variety 'Trishna' having high geraniol content (93.6%).

Somaclonal variants generated through tissue culture can also led to the development of a new culture for specific requirements *viz.*, tolerance to salinity-alkalinity etc.

iii. **Hybridization (intra- and inter-specific):** Keeping in view the viable and profuse seed setting, the technique of hybridization among improved and diversified clones/types is important to generate diversity. Similarly, wide crosses are also desirable to transfer resistance against biotic and abiotic stresses from related species. The inbred lines developed by continuous selfing can be crossed among themselves after evaluating for general combining ability (GCA) and the heterosis (hybrid vigour) thus obtained can be exploited by taking advantage of vegetative propagation in these crops. Sobti *et al.*, 1978 made an inter-specific cross between *C. nardus* var. confertiflorus x *C. jawarancausa* at RRL, Jammu and selected a line called 'Jamrosa', with 54-83 percent geraniol and 17.89 percent geranyl acetate having hardiness and vigorous growth.

However, its oil was inferior in odour to palmarosa oil due to presence of piperitone, myrcene and b-ocimene content.

iv. **Polyploidy:** Keeping in view of the somatic chromosomes from diploid to hexaploid in the genus *Cymbopogon*, and importance of vegetative parts, efforts should be made to create polyploidy in crop by using colchicine and by other means to utilize variants thus created by doubling the chromosome number or by combining chromosome sets from different species in their amphidiploids.

v. **Population improvement method:** Recurrent selection is very useful method for population improvement and it is practiced in a synthetic variety or composite population to improve yield and quality by concentration of additive variances.

Varietal Achievements

The following crop varieties of palmarosa have been developed from various institutes utilizing different breeding approaches.

Variety	Name of Institute	Characterization
RH-49	CCS HAU, Hisar	Clonal selection, Geraniol (>90%)
IW-31243	NBPGR, New Delhi	-
IW-31245	NBPGR, New Delhi	Geraniol (90%)
Trishna	CIMAP, Lucknow	Polycross variety Geraniol (93%)
Tripta	CIMAP, Lucknow	Geraniol (94%)
PRC-1	CIMAP, Lucknow	Geraniol (93%)
CI 80-68	JNKV, Indore	Geraniol
CIM-Harsh	CIMAP, Lucknow	Geraniol (94%)
Vaishnavi	-do-	
IW-3629	NBPGR, New Delhi	
IW-3244	NBPGR, New Delhi	
OPD-1	-	
OPD-2	-	
RRL(B)-71	Bhuvneshwer	
RRL(B)-77	-do-	
Jam Rosa (RRL-82)	Jammu	Interspecific hybrid variety

References

Baradu, B.L. and C.K. Atal. 1974. Indian Perfumer, 18(1): 7.

Gupta, R., K.C. Trivedi, B.G. Joshi, S.K. Pareek and M.L. Maheshwari. 1980. Indian Perfumer, 24(3): 134-41.

Pareek, S.K., D.B. Sexena, M.L. Maheshwari and R. Gupta. 1981b. Indian Perfumer, 25(3&4): 64-70.

Pareek, S.K., M.L. Maheshwari and R. Gupta. 1981a. Indian Fmg., 36(4): 37-40.

Punia, M.S., P.K. Verma and G.D. Sharma. 1986. Indian Perfumer, 30(1): 240-246.

Sahoo, S. and P.K. Dutta. 1982. Indian Perfumer, 26(2-4): 41-45.

Sethi, K.L., M.L. Maheswari and R. Gupta. 1986. Indian Perfumer, 30(2-3): 309-313.

Sharma, J.R., H.O. Mishra and R.K. Lal. 1987. Bafai. J., 9(2): 21-26.

Sobti, S.N., B.L. Rao and B.B. Jain. 1978. Indian perfumer, 22(4): 281-83.

Verma, P.K., G.D. Sharma, M.S. Punia, J.S. Hooda and G. Talwar. 1990. J. Econ. Bot. and Phytochem., 1: 7-10.

Yadav, O.P., I.S. Yadav, V.K. Madan, P.K. Verma and M.K. Deen. 2012. Abstracted in Proceedings of Seminar on "New perspectives in Aromatic and Medicinal Plants" held at CCS HAU, Hisar on Feb. 8-9, 2012, p.146.

Suggested questions

1. What are the two types of Palmarosa?
2. To which family Palmarosa belongs.
3. What is the pedigree of an interspecific hybrid variety of Palmarosa?
4. What is the ploidy level of two types of Palmarosa & their somatic chromosome numbers?
5. Which part of Palmarosa has the maximum oil content & which part has the best quality of essential oil.
6. Which is the best stage of harvest in Palmarosa to have high oil yield.
7. What is the ideal plant type/best ideotype of Palmarosa?
8. Name the varieties developed by CIMAP, Lucknow.
9. What are the effective breeding methodology and why?
10. What are different uses of its essential oil?

20

Patchouli (*Pogostemon cablin* Benth.)

Introduction

Patchouli is known as *patcholi* in Hindi, *patche tene* in annada, *patchapan* or *patcha* in Marathi, *patchilla* in Malayalam, *pacchilai* in Tamil. It is a hardy perennial herb well-adjusted against the hot and humid climatic situations.

Patchouli is an aromatic herb of immense importance due to presence of essential oil extracted from shade dried leaves. It possessed a powerful sweetness and spicy fragrance. Patchouli has tremendous business potential owning to its unique fragrance properties in the global market. Its oil easily blends with other essential oils giving strong base as fixative properties in preventing evaporation and promoting tenacity. Worldwide, Indonesia is considered as the chief patchouli oil producer.

Origin

Pogostemon cablin Benth. is natural to Philippines but also found growing wild in many countries of South Asian region. The word 'cablin' is derived from 'cabalam', a local name in the Philippines. Presently, it is not only cultivated on commercial scale in India but also in Indonesia, Singapore, China, Malaysia, Brazil, West Africa, Seychelles, Vietnam and West Indies.

Botany

It belongs to family Lamiaceae and grows up to 1.2 m height having broad leaves and erect stem. Its plant bears small pale to pink-white flowers. The leaves margins are lobed and with abundant hairs present on dorsal surface of leaf, well-known to collect essential oil in the glandular trichomes. Plant is multiplied through asexual mode of propagation i.e. shoot tip cuttings as its plants have sparse flowering and failure in seed set also observed.

Genotypic variation and Cytogenetics

The genus *Pogostemon* has about 40 species and out of which 20 are found in India. The chromosome numbers of some of the species are as under.

Species	Somatic chromosomes (2n)	References
Pogostemon cablin Benth.	34	Lavania, 1984
P. heyneanus Benth. (*Syn. P. fleurissant* Benth)	32	Indiramma Basavaraj, 1982
P. paniculatus Benth.	32	-do-
P. plectranthoides Desf. (*Syn. P. benghalensis*)	32	Vij and Kashyap, 1976
P. parviflorus Benth.	32	Gill, 1971
P. purpurascens	32	Indiramma Basavaraj, 1982
P. speciosus	34	-do-
P. wightii	32, 34	-do-
P. hortensis	-	-
P. comosum Miq.	-	-
P. tomentosus Hassk.	-	-
P. pubescence	-	-
P. heyeanus	-	-

Genetic variability

There is limited the gene pool and not much of genetic variability is availability in the crop as such. Hence initially five exotic strains were collected and evaluated for their herbage and oil yield potential under south Indian conditions. It was first introduced from Singapore in 1935 as living plants/cuttings and named as **'Singapore'**; and identified as *P. cablin* Benth.

Subsequently, another strain introduced in 1953 was named as **'Johore'**; and **'Indonesian'** introduced in 1954. Later on, **'Malaysian'** and **'Java'** strains were also introduced and grown at Indian Institute of Horticultural Research (IIHR), Bangalore (Nirody *et al.,* 1961).

Evaluation of above listed strains at IIHR, Bangalore exhibited enough variation for their leaf morphology, trichome density and plant canopy. However, the quality of oil obtained from **'Johore'** was found to be superior over the **'Indonesian'** and **'Singapore'** strains.

On the other hand, strains **'Java'** and **'Singapore'** were found to flower under Bangalore conditions, but with no seed setting. Therefore, no hybridization and genetic improvement works could be undertaken.

Thereafter, efforts were made to create genetic variability through **mutation** and to develop **polyploid** clones for development of higher herbage yield, oil yield and oil quality in addition resistance to nematodes (Nirody *et al.*, 1961).

Later on, it was further reported that strains **'Johor'**, **'Indonesian'** and **'Malaysian'** had higher herb yield potential, besides superior and comparable oil quality to that of imported oils in terms of their chemical composition and odour value (Kumar *et al.*, 1989).

However, the work is still in progress to develop suitable agronomic practices for higher herb and oil yield at Indira Gandhi Krishi Vishwavidyalaya at Raipur (Chattisgarh).

Varietal development

CIM-Shreshtha: It is the first variety released by CIMAP, Lucknow, which is tolerant to diseases and having oil yield potential of 80 -100 kg/ha.

Uses

- Used to improve soils with dampness, exterior syndrome and relieve summer heat
- "Pogostemoni Herba" is widely used in the hygiene and cosmetic industries.
- Used for treatment of diarrhoea, abdominal pain, colds, headaches, nausea, vomiting, and snake & insect bites.
- Essential oil is used to make the body lotions, detergents soaps and perfumes.
- Its oil is also used in aromatherapy to control appetite, improve sexual interest. relieve depression, stress and calm nerves.
- Oil has insecticidal, anti-thrombotic, aphrodisiac, anti-depressant, anti-bacterial, anti-fungal, anti-microbial, anti-oxidant, analgesic, anti-inflammatory, anti-platelet, anti-mutagenic, anti-emetic, fibrinolytic and cytotoxic properties.

Chemical Composition

Recent studies have been focused on isolating individual compounds to understand their mechanism involved in various pharmacological activities. The plant contains various phytochemicals, including phytosterols, flavonoids, organic acids, lignins, glycosides, alcohols, monoterpenoids, triterpenoids, sesquiterpenoids and aldehydes.

Crop Production

Climate: It best suited to cultivate under tropical conditions having rainfall from 1500-3000 mm, high temperature ranging between 25-35 °C and high relative humidity. It can be grown under open or partially shaded conditions upto 50% shade.

Nursery raising: It is planted by rooted stem cuttings raised in the nursery buds under shade. 10-12 cm long terminal cuttings are planted on raised beds or in polythene bags.

Soil requirement: Wide range of soil conditions are suited, while light porous, well-drained soils which are rich in organic matter and having pH range between 5.5-6.5.

Planting time: Plantation of rooted cuttings of 40-45 days can be done during rainy season in *kharif* on raised beds/ridges at spacing of 45×45 cm, avoiding hot and sunny conditions. Suitable raised beds are prepared in the field with provisions for irrigation and drainage. Light irrigation is necessary soon after transplantation, and gap-filling should be done as and when necessary.

Fertilizer requirement: Soils with medium fertility are applied with 10 tons of farm yard manure (FYM) along with chemical fertilizers of 100 kg N, 50 kg P and 50 kg K per hectare. Nitrogen is split in 3 monthly doses, whereas entire dose of P and K is applied while preparing field.

Inter-culture operations: Keep field weed-free in the initial stages for good growth, as a fully grown crop competes well with the weeds in later stage.

Irrigations: Irrigations are given in low-rainfall areas as per need. Weekly irrigations are required either by surface, sprinkler, or drip methods in non-rainy periods; but very sensitive to water logging conditions. In the ridge and furrow method of planting, alternate furrow irrigation saves half of the water requirement.

Harvesting stage: It should be harvested when turns to light greenish-yellow colour from thick green with a mild fragrance. It takes 4-5 months for first harvest after planting; and subsequent harvests can be taken at 3 month intervals. Leaves and tender twigs are cut at reasonable height for better regeneration of the crop with sharp sickles or secateurs to avoid disturbance to the root system. Avoid harvesting of stem portion and to have 60-70% leaves for good oil recovery.

Fresh herb yield: Its production to the tune of 150-250 quintals per hectare can be obtained under optimum conditions from two harvests. Air-dried herb yield of 30-50 quintals can result in 60-100 kg of oil per hectare.

Oil storage: The oil should be filled in suitable containers (preferably aluminium), air-tight and light proof containers and stored in a cool place.

Plant protection

Diseases: Leaf blight (*Cercospora* sp.), Wilt (*Rhizoctonia solani*); Viral diseases; and Yellow mosaic caused by virus transmitted by white fly (*Bemisia tabaci*) are prevalent under suitable conditions.

Insects: Leaf feeding caterpillar (*Pronomis profusalis* Warren.); leaf roller (*Herpetogramma licarsisalis*) and myriad bug (*Pachypeltis* spp.) are prominent insects.

Nematodes: Root-knot caused by *Meloidogyne incognita* and *M. hapla.may* occur.

References

Gill, L. S. 1971. Caryologia, 24(2): 203.207.

Indiramma Basavaraj. 1982. Chemotaxonomical and karyological studies in Labiatae. Ph. D. Thesis, Department of Botany, Bangalore University, Bangalore.

Lavania, U.C. 1984. Taxon, 33: 757.

Nirody, B. S., Chacko, M. C. and Bhati, A. 1961. Indian Perfumer, 5(2): 75-84.

Vasantha Kumar, T., Maheshwari, M. L., Chandravadana, M. V. and Surekha, N.1989. 11th Intern. Cong. Essen. Oil, 1: 119.

Vij, S. P. and Kashyap, S. K. 1976. Cytologia, 41: 713-719.

Suggested questions

1. What is the scientific name of Patchouli?
2. To which family Patchouli belongs.
3. What is somatic chromosome number of Patchouli?
4. What is the best harvesting stage of Patchouli?
5. Name the variety developed by CIMAP, Lucknow.
6. What are various phytochemicals of its essential oil?

21

Damask Rose (*Rosa damascena* Mill.)

Introduction

Among flowers, the rose enjoys the greatest adorable position over the world. In real sense, rose almost hypnotised the man most by its colour and fragrance. Apart from its aesthetic value, the roses are of high economic benefits, and religious significant. Besides, it finds it place in art and culture from times immemorial. The rose thus merits pride of place as the **'Queen of flowers'**.

Origin

The damask rose is originated in ancient Persia (now Iran), and later on spreading over Europe and Northern Africa. Further, widespread farming of *R. damascena* for rose oil manufacturing began in Turkey and Bulgaria. Many countries of the world grow scented roses for extraction of the volatile essential oil. The genus *Rosa* grows wild in many parts of the temperate and sub-temperate regions. The major growing countries are Bulgaria, France, Italy, Turkey, Russia, China and India. Bulgaria is the principle European country for supply of rose oil to the world market. Among different species, *Rosa centifolia* Linn. is largely grown in France and Italy, whereas *Rosa damascena* Mill. in Turkey, Bulgaria, Russia and Persia. *Rosa damascena* Mill. was first introduced successfully in middle east countries during 1680, resulting in the establishment of entire area as **'Valley of flowers'**. It belongs to family rosaceae.

Though introduction of damask rose (*Rosa damascena*) in India is not clear, yet it obtained adorable position among royal palaces and hermitages of saints. However, the establishment of rose oil and rose water industry in India continued for the last over 300 years.

Distribution

Presently, Bulgaria is the world's biggest producer of damask rose oil, accounting for over 80 percent of global output (15–20 tonnes) in its provinces. Turkey produces the remaining 15 percent of rose oil, while the remaining countries accounting for only 5 percent of total production.

In India, the production of damask rose is centred in Uttar Pradesh, Rajasthan, Jammu and Kashmir, Himachal Pradesh, some sections of Punjab, Uttarakhand, and hill areas of Tamil Nadu and Kerala, covering an area of about 2500–3000 ha.

The market for damask roses is growing year after year. The current global rose oil production accounts for about half of the total demand primarily due to the low oil content, short flowering season, and cost-intensive production (Pal, 2013).

Genotypic variation

Genus *Rosa* has more than 200 species, but only a few of them are being utilized for essential oil production. *Rosa damascena* Mill. called as damask rose is one of the best known scented rose species grown worldwide for the production of rose essential oil. It belongs to family Rosaceae.

The following three species are commercially valuable among the oil bearing roses.

1. ***Rosa centifolia*** Linn.**:** This is considered to be native of eastern Caucasus. Earlier, it was cultivated extensively for extraction of rose attar, but now *R. alba* L. is grown in gardens for ornamental purpose only and no commercial utilization.
2. ***Rosa bourboniana*** Desp.**:** This is commonly known as 'Desi Gulab' and 'Baramasi' or 'Edward Rose'. rose oil and attar. But the oil content in the fresh petals is very low (0.013-0.015 percent), and hence not used for production of rose attar.

   It is widely cultivated in Kannuaj, Kanpur, Lucknow, Deogarh, Madhupur, Calcutta, Pushkar (Rajasthan) and in gardens of North India. In Southern India, it is grown in districts of Madras, Chinglepat, Tanjore, Trichnopally, Coimbatore; Madurai of T.N.; and Chenepatra, Nanzaigud and Bangalore in Karnataka. Tanjore was well known for the production of rose water and its quality.
3. **Rosa *damascena Mill.*:** It is commonly known as 'Damask Rose' and surpasses all over roses in fragrance. Hence used for commercially for production of oil and attar. Its plant is bushy and hardy with long life

span (20-30 years). It is commercially grown in Aligarh, Ghazipur, Balia (U.P.) and Haldighatti area in Rajasthan.

Cytogenetics

Rosa damascena is regarded to be of hybrid origin. Genetic analysis has shown that there are two types of damask rose, namely, *Rosa damascena* (*R. galbia x R. phoenicea*) and *Rosa bifera* (*R. galbia x R. moschata* or *R. canina*).

The plant is thought to have a tri-parental origin that includes two consecutive crosses (*R. moschata* x *R. gallica*) x *R. fedschenkoana* Regel and *R. moschata* as the maternal ancestor (Iwata *et al*., 2000).

Cultivated *Rosa damascena* is a complex allo-tetraploid with a stable chromosome number of $2n = 4x = 28$ (Gudin, 2000). Most of the *Rosa* species were found to be tetraploid ($2n = 28$) with basic chromosome number $x = 7$. Besides, triploid ($2n = 21$) was also observed (*R. centifolia*); and pentaploid ($2n=35$) in Iranian cultivars.

Chemical composition

In damask rose oil, about 300 volatile constituents have been recognised. It belongs to a diverse group of compounds, including monoterpenes, long-chain hydrocarbons, and metabolites from carotenoid degradation (Nunes and Miguel, 2017). However, the most common compounds include geraniol, citronellol, linalool, and nerol, which found 60 percent of its oil content.

Kahol and Agarwal (1983) reported overall chemical compositions of rose oil produced in India as below.

Oil constituents	**Quantity (%)**
Geraniol	34.9
Citronellol	23.9
Nerol	12.4
Phenyl ethyl alcohol	7.4
Linalool	2.30
Eugenol	1.60
Rose oxide	1.30
Ethanol	1.20

Factors affecting oil quality

Besides genotype, there are many factors that also play a major role in determining the quality of rose oil. It depends on propagation method, growth conditions, harvesting procedures, method of distillation, time and level of pruning, time of flowers harvesting, storage of plant material, and prevailing climatic conditions.

Breeding methodology

The most prevalent method used to improve damask rose varieties is traditional breeding method of clonal selection. Plant breeders employ morphological variances to pick superior progeny, but this is a time-consuming and sometimes insufficient method because many species have a huge and overlapping range of morphological variations that are affected by climatic factors.

i. **Clonal Selection:** Patra *et al.* (1987) initially selected 35 variables out of original population of 576 plants and finally selected four clones at CIMAP, Lucknow. These were evaluated along with two other clones (RL-19 and Bulgarian rose) for chemical composition of oil. Clone RSL-31 was found the best over the Bulgarian rose having high rhodinol, geraniol, nerol, ethanol and rose oxide as given below:

Oil composition	**Genotypes**					
	RSL-31	**RSL-76**	**RSL-87**	**RSL-88**	**RSL-19**	**Bulgarian rose**
Rhodinol	53.80	53.42	75.03	63.05	45.70	35.00
Geraniol	18.89	6.02	7.78	10.54	8.27	18.40
Nerol	6.69	2.96	3.67	6.31	3.60	5.90
Ethanol	3.38	0.81	0.50	3.0	2.13	1.43
Linalool	1.91	0.80	0.97	1.33	1.25	2.18
Rose oxide	0.81	0.68	0.91	0.83	0.62	0.43
Phenyl ethyl	1.02	1.31	1.90	1.22	0.90	1.45
C. 19 Paraffin	-	17.75	-	-	20.16	14.50
C. 23 Paraffin	-	6.03	-	-	8.49	4.20

From the morpho-physiology, the plant of **'Ranisahiba'** was distinct from its mother variety **'Kannauj'** control, as well as two controls i.e. **'Noorjahan'** and **'Aligarh'** land races.

The novelty in invention of 'Ranisahiba', besides being superior to the existing rose varieties and land races for flower yield, distinctly out-classed all existing rose genotypes/land races by having high oil content, oil yield and high geraniol content coupled with least concentration of undesirable wax component in the essential oil.

The variety **'Ranisahiba'** is a perennial shrub with upright prickly shoots. The synchronous branching of shoots gave a globular canopy of the plant. Globular canopy allowed equal distribution of sunlight to the flowers, leading to enhanced flower biomass. The flowers bloom largely synchronously leading to brevity in flower picking. 'Ranisahiba' was assessed for its stability for five years at two contrast climatic conditions of sub-tropical plain (Pantnagar) and temperate hill (Purara). The results obtained revealed its high stability and adaptation over the environments.

'Ranisahiba' being vegetative propagated by stem cuttings, its morpho-physiological traits, including oil yield are firmly fixed. The plant of 'Ranisahiba' was frost sensitive during chilling winter (0-5° C) of December and January at higher altitudes (>500 m). To overcome the hostility of the chilling winter, the plants needs deep pruning during first week of December. The mature plant parts after pruning are resistant to frosting and start sprouting during mid-January after a brief period of remaining dormant at both hill and foot hill situations. However, it is worthwhile to mention that sprouting of shoots as well as flower bud formation are early by 15 to 20 days as compared to the normal cultivars. The flower buds in profuse number were observed at the foot hill area during first week of March, as against that during last week of March in variety 'Noorjahan' and others. The early sprouting behaviour in particular was a major physiological 'marker' to identify some winter hardiness in the new genotype 'Ranisahiba'.

ii. **Biotechnological methods:** With the development of new tools and technology *viz.* genome sequence, SNP array, genetic mapping, NEXT-generation sequencing, etc. have been extensively utilized to identify the QTLs in *R. damascena* related with desired traits. DNA markers and their transferability to *R. damascena* from other species provided a new opportunity to use molecular marker-assisted selection (MAS), having potential to improve the effectiveness in selection. The development of transcriptome data also aided by next-generation sequencing. Organizing this information and linking it to specific phenotypes will help us learn more about molecular mechanisms and genetics.

Today, for detecting the polymorphism regarding to know about the genetic diversity among rose cultivated varieties, the Microsatellite markers were also found to be highly informative. The dendrograms made by using molecular data clearly differenced among the cultivars based on their dissimilarities and also make groups based on genetic similarities. The SSRs also used in the rose study were found helpful for proper identification of rose cultivars. This technique also protects the cultivars against unauthorised use for commercialization and also for documentation of diversity from the germplasm conservation of Indian rose.

Varietal development

Following varieties have been developed by various institutes.

Variety	Institute	Characters
Noorjahan	CIMAP, Lucknow	Oil content (0.04 %), Citronellol (20.8 %), Geraniol (25.3 %)
Ranisahiba	-do-	Oil content (0.8 %), Geraniol (30-35 %), Geranyl acetate (5.4-7.0 %), Citronellol (4.6-5.0 %), PE Alcohol (7-8 %), Linalool (1.1-2.5 %)
Bulgarian rose	Introduction	Oil content (0.02 %), Geraniol (18 %), Geranyl acetate (1.6 %), Citronellol (33.3 %), PE Alcohol (1.4 %), Linalool (2.2 %)
Arka Pride	IIHR, Bangalore	Cut flower variety suitable for protected cultivation.
Arka Sukanya	-do-	Fragrant variety (0.22% oil). Floriferous nature and tolerant to pest and diseases. Ideal choice for garden display.
Arka Ivory	-do-	Ideal for cut flower production under protected cultivation. Moderately resistant to mites.
Jawala	IHBT, Palampur	Suitable for subtropical northern plains, low hills to mild temperate regions (up to 1200 m). Flowers in March-April and September. Tolerant to hailstorms, rains and high velocity winds. Bears short clusters of flowers in compact bunches.
Himroz	-do-	Ideal for mild temperate to cold temperate regions (1200-2500 m). Winter tolerant without any visual sign of injury. Flowers for a period of 22-25 days during the early summer.
Arka Sinchana	IIHR, Banglore	Highly floriferours ever blooming floribunda rose identified for garden display. Flowers are bright red in color. Flowers are small in size (5.4 cm), and filled with multiple petals.
Arka Sharmeeli	-do-	Ever blooming floriferous Hybrid T rose. Flower color changes from light pink to different shades of red from bud to different stages of blooming. Flowers are medium size with an average flower diameter of 7-8 cm.
Arka Kinnari	-do-	Ever blooming floriferous Hybrid T rose. It produces bicolor flower consisting vermilion red shading towards orient pink. Characterized by long stalk with cluster of flowers. It has dark green shining foliage. Flowers are large with an average flower diameter of 10-11cm.

Arka Savi	-do-	Identified for commercial cultivation for loose flower production. Flowers are purple pink and produced in bunches. The rose hybrid is identified for its floriferous nature with high yield. In addition to high yield, it has added advantage of long shelf life (5 -6 days).
Arka Swadesh	-do-	Identified for cut flower under polyhouse. Long stalk flowers with stalk length of 65-70 cm. High yielding capacity with medium sized flowers and good keeping quality (6 days). Shining foliage and flowers of attractive red colour.
Arka Parimala	-do-	Red-purple coloured flowers are fragrant. Moderately resistant to thrips and black spot. High flower yield potential.

Uses

- There is a broad range of industrial applications of rose essential oil in perfumes and cosmetics, flavouring tobacco, liquors, sweet dishes in the food industry and pharmaceutical industry.
- Indian System of Medicine used damask rose preparations as astringents, tonics, mild laxatives, and anti-bacterial in the treatment of sore throats, swollen tonsils, heart issues, eye illness, and gall stones.
- The natural products isolated have important biological properties, such as cytotoxic, anti-mutagenic, anti-viral, anti-microbial, anti-oxidant, and anti-neoplastic activities.

Other products from Damask Rose

1. **Rose water:** Major item of production in India, as more than 80 percent of flowers are used for production of rose water. It owes its scent mainly to phenyl ethyl alcohol. Produced during flowering season by hydro-distillation of rose flowers and stored for 2-3 months for maturation of odour. Later on diluted with distilled in different proportions and stored.
2. **Rose attar:** Obtained by distilling flowers and distillate collected over oil of sandal wood. For cheaper products, liquid paraffin is used in place of sandal wood oil. Largely used as perfume in agarbatties and flavouring tobacco for snuffs and chewing purpose.
3. **Gulkand:** By-product of conservation of flower petals with sugar in 1:2 ratios, kept in sun for about 30 days for maturation. Very good tonic having laxative properties.
4. **Gul-roghan:** Produced by maceration of rose petals with warm sesame seed oil and used as hair oil.

5. **Punkhuri:** Shade dried rose petals used for preparation of cool summer drink (sharbat).
6. **Otto of Rose:** Also known as rose oil or 'Ruhe Gulab'. Prepared by distillation of rose flowers and distilling the same 2-3 times until saturated with the oil dissolved in it. Later on, chilled and the oil drops floating on the surface of water are removed. Generally, its quantity is extremely low (0.0045 %) on fresh weight basis and hence very expensive in the inter-national market.

Genotypic variability for flower colour

References

Chandra, V. 1995. In: Advances in Horticulture. Vol. 11. Medicinal and Aromatic Plants (1995). Edited by K.L. Chadha and Rajendra Gupta, Malhotra Publishing House, New Delhi, India, pp. 843-858.

Kahol, A.P. and Agarwal, K.K. 1983. Indian Perfumer, 37(2): 137-142.

Patra, N.K., Srivastava, S.K., Srivastava, K.K. and Naqvi, A.A. 1987. Indian J. Agric. Sci., 57(12): 938-940.

Suggested questions

1. What is the scientific name of Damask Rose?
2. To which family Damask Rose belongs.
3. What is somatic (2n) and genomic (x) chromosome numbers of Damask Rose?
4. What are the various products made from Damask Rose?
5. Name the best variety developed by CIMAP, Lucknow and its oil content.
6. What are the three species of Damask Rose?
7. What are major chemical constituents of its essential oil?

22

Tulsi/Basil (*Ocimum* species)

Introduction

Basil/Tulsi is a popular name given to any aromatic herb that belongs to genus *Ocimum* and family Lamiaceae (Labiatae). The word 'basil' was derived from Greek word 'Basilica' which means the royal plant. It comprises over 160 species spread over the tropical, sub-tropical, and warmer part of temperate regions of both the hemisphere ranging from sea level to 2000 m height with their diversity met in West Africa. Tropical Africa has maximum diversity with 59 species, followed by sub-tropical region of South Africa with 19 species and India with 9 species only.

Distribution of different *Ocimum* species

***Ocimum* Species**	**Distribution**
O. basilicum and *O. sanctum*	Entire Indian sub-continent
O. americanum, O. canum	North-western India (J&K, Punjab, H.P., Delhi and U.P.)
O. suave, O. viride *O. carnosum*	Whole of Deccan plateau and coastal region of southern-eastern parts of Indian peninsula (Introduced in 1972)
O. gratissimum	Southern-eastern part of India
O. adscendens Willd.	South India
O. killimandscharicum	Dehradun (Utranchal), Lucknow (U.P.), West Bengal and Kashmir (Introduced from Kenya)

These different species have mainly methyl chavicol, linalool, methyl cinamate, camphor and eugenol as chemical constituents in different quantities for making it of different quality and odour.

Uses

The essential oil obtained mainly from leaves and tender parts of shoots have immense value as flavour as well as fragrance, in medicine perfumery and in flavouring foods. Besides, the above conventional uses, the natural essential oil especially from *Ocimum* species have the potentiality of being very safe insect controlling agents. The oil also finds wide applications in confectionary, foods, condiments, dental cream and mouth washes etc. It is used in perfuming, flavouring and pharmaceutical product.

Genotypic variability under *Ocimum* species

Common Name	Scientific Name	Oil Quality parameters
Sweet/Indian/ French basil	*Ocimum basilicum*	Linalool: 40%, Methyl chavicol: 35%
Holy basil	*Ocimum sanctum*	Eugenol
Hoasy basil	*Ocimum viride*	Thymol: 65-70%
Mint basil	*Ocimum canum*	Linalool: 70-80%
Long Tulsi	*Ocimum gratissimum var. clocimum*	Eugenol: 75-80%
Long Tulsi	*Ocimum americanum*	Methyl chavicol: 70-75%
Camphor/Kapoor basil	*Ocimum killimandscharicum*	Citral:75-80% Camphor

Ocimum sanctum var. Rama.

Ocimum sanctum var. Shyama

Ocimum basilicum L.

Ocimum gratissimum var. clocimum

Cytogenetics

Cytological studies provided valuable information regarding inter-relationship and evolutionary trends in genus *Ocimum*. A survey of chromosome number recorded in this genus, clearly suggested the existence of more than two basic numbers. The different haploid (n) chromosome number so far recorded are

12, 13, 16, 20, 24, 32, 36 and 38, indicating clearly that the aneuploidy and polyploidy have played an important role in the diversification of this genus.

***Ocimum* species**	**Chromosome No. (2n)**	**References**
***Basilicum* group: x=12**		
Ocimum canum Sims.	24 (Diploid), 26, 64	Pushpangadan *et al.*, 1975; Getsadze, 1975
Ocimum basilicum L.	48 (Tetraploid)	Verma, 1947 Pushpangadan *et al.*, 1975; Getsadze, 1975
Ocimum americanum L. *(O. basilicum* x *O. canum)*	72 (Hexaploid)	Sobti et al., 1975
Ocimum killimandscharicum Guerke.	76	Mehra and Gill, 1972 Pushpangadan *et al.*, 1975
***Sanctum* group: x=8**		
Ocimum sanctum (Purple & Green)	32 (Tetraploid)	Mehra and Gill, 1972
Ocimum gratissimum L.	40 (Pentaploid)	Khosla and Sobti, 1985 Getsadze, 1975
Ocimum carnosum	48 (Hexaploid)	Khosla and Sobti, 1985 Singh, 1978
Ocimum viride Willd.	40 (Pentaploid)	Singh, 1978
Ocimum sauve Willd.	64 (Octaploid)	-

Cyto-taxonomical Investigations

Pushpangadan (1975) and Khosla (1985) have broadly classified the genus *Ocimum* into two groups i.e. *basilicum* group and *sanctum* group on the basis of comprehensive study on morphology, cytogenetics, breeding, chemical characteristics etc. of the species available in India.

***Basilicum* group**	***Sanctum* group**
Basic chromosome No. x=12	**Basic chromosome No. x=8**
Herbaceous; Flower more conspicuous	Biennial, triennial or perennial under shrub; Flower less conspicuous
Seeds are black, ellipsoid and mucilaginous (on wetting)	Seeds are brownish, globose, shining and non-mucilaginous

It was further suggested that polyploidy species in the genus *Ocimum* have most probably originated as allo-polyploids. The floral structure of *Ocimum* species is most suitable for pollination by insects. A comparative study of seed set obtained by selfed and open pollinated plants has indicated that cross pollination is more common, although the plants are not completely self-incompatible. Good seed set by selfing is obtained in the species belonging to *sanctum* group. The *Ocimum* species are basically cross pollinated. Therefore, most of the species thus maintain a certain amount of genetic variability in the population.

Comparison of two groups of *Ocimum* species

Species	Major chemical constituents
***Basilicum* group: x=12**	
O. canum Sims.	Linalool or camphor
O. basilicum L.	Methyl chavicol, methyl cinnamate, eugenol, linalool
O. americanum L.	Methyl chavicol, citral
O. killimandscharicum Guerke	Camphor
***Sanctum* group: x=8**	
O. sanctum L.	Eugenol
O. gratissimum L.	Eugenol
O. viride Willd.	Thymol
O. suave Willd.	-

Genetic variability

Verma *et al*., (1989) reported that different species of *Ocimum* varied greatly among themselves for different morphological traits.

S. No.	Traits	Species
1	Leaf + Inflorescence/ Stem ratio	*O. sanctum* (1.55) *O. citriodrum*, (EC 110586-3.30)
2	Oil content (%, FWB)	*O. basilicum* (French basil), (EC 112809-0.16%) *O. basilicum* (Indian basil), 0.43%
3	Oil yield (lt./ha)	*O. basilicum* (Indian basil), 280.6 lt./ha *O. basilicum* (French basil), 229.4 lt./ha

Variability parameters, heritability (broad sense) and genetic gain for different characters in different *Ocimum* species

Parameters	Plant height (cm)	Fresh herb yield (q/ha)	Oil content (%, FWB)	Oil yield (lt./ha)
Range	48.0-63.7	97.02-415.1	0.21-0.51	43.0-160.4
GCV (%)	10.68	35.22	24.55	41.34
PCV (%)	11.55	35.66	24.99	41.84
Heritability % (BS)	85.60	98.21	96.84	92.81
Genetic gain (%)	20.36	71.90	49.86	84.14

Hybridization

Floral organs in all the species of *Ocimum* studied are identical. The flowers are prot-androus (anthers mature first), but duration of time between opening of flowers and dehiscence of anthers and maturity of style varies from one species to other species. Flower structure of *Ocimum* species is most suitable for cross pollination by insects, mostly bees. A comparative study of seed set obtained from selfed and open pollinated plants indicated that cross pollination is more common, although the plants are not completely self-incompatible.

Inter-specific hybridization

***Sanctum* group (x=8):** Good seed-set by selfing is obtained in species belonging to which have short-lived flowers. There is strong tendency in the species to out-breed within the population. Although cross breeding within the population or between the types or races are common in this group, but inter-specific crossing is rare.

***Basilicum* group (x=12):** The frequency of inter-varietal hybrids varied from 5.8 to 18.5 per cent only. Inter-specific crossing and production of hybrids among species (*O. canum, O. americanum, O. basilicum and O. killimandscharicum*) of this group do occur when they are brought together. These hybrids are either completely sterile or with partial seed setting. The seed obtained from these partial F_1 inter-specific hybrids exhibited great variations in morphological, cytological and chemical characteristics of the plants.

Snegirev (1936) made inter-specific cross between species of the above two group *viz.*, *O. canum* x *O. gratissimum* and studied the oil content of the hybrid. He reported that camphor content of one hybrid increased from 5-27 per cent; and eugenol content from 5 to 28 per cent. He further reported that higher content of camphor was usually associated with higher content of eugenol. On the other hand, another hybrid contained more than twice eugenol in comparison to the parent (*O. gratissimum*), but the camphor content was never higher than that of the other parent (*O. canum*).

Some of the amphidiploids synthesized in the case of sterile F_1 inter-specific hybrids involving different species have also been found useful from the essential point of view.

Therefore, hybridization in *Ocimum* species and other aromatic plants could be of great value in the production of volatile oils of great industrial value.

Mutation breeding

Results of a few work done on mutation breeding are as below

***Ocimum* species**	**Mutagen used**	**Character reported**	**Reference**
O. basilicum L.	X-rays	High oil content	Buzinov and Zhir, 1962
O. gratissimum L.	Gama Irradiation	High oil content and increased resistance to *Fusarium* wilt.	Getsadze and Kankaya, 1976

Inheritance of chemical constituents of oil

Pushpangadan *et al.,* (1975) studied the difference in essential oil in genus *Ocimum* showed that citral, linalool and geraniol (which are monoterpenes) are inherited independently of methyl chavicol and eugenol (which are phenolic in nature).

Three chemical races *viz.* rich in camphor, eugenol and methyl chavicol have been isolated from *O. basilicum* var. glabratum (French basil). They showed that the presence and absence of these constituents in French basil is controlled by a single gene, which exists in three alleles. Pushpangadan and Sobti (1975) have sown that the gene responsible for citral as dominant over gene responsible for geraniol and linalool.

Breeding Strategies

Ocimum species in general, like, medicinal and aromatic plants have very short history of cultivation and direct artificial selection. *O basilicum* var. glabratum (French basil), used mainly in Europe and America, is perhaps the only species of *Ocimum* that has been cultivated as commercial crop.

The method suited for breeding better species of *Ocimum* species are:

a) Population improvement approach and identification of promising plants.
b) Study on general and specific combining ability of desirable traits for exploitation of heterosis.
c) Recurrent selection and polycross of selected lines to produce synthetics.

Varietal achievements and chemical constituents in different *Ocimum* species

A wide array of plant types, rich in linalool, citral, methyl chavicol, thymol and eugenol were evolved. Some of the important cultivars/hybrid varieties released for cultivation from RRL, Jammu are given below:

Average yield data along with quality of selected and improved strains of *Ocimum* species

Ocimum species	Fresh herb yield (q/ha)	Oil yield (lt./ha)	Oil content (%, FWB)	Major oil constituents
O. canum, RRL-OC-1 (Selection)	400	200	0.50	Linalool (70-72%)
O. americanum, RRL-02 (Selection)	410	185	0.45	Methyl chavicol (70-75%)
O. americanum, RRL-07 (Improved strain)	400	200	0.50	Citral (75-80%)

O. gratissimum, RRL-08 (Improved strain Clocimum)	400	160	0.40	Eugenol (75-80%)
O. viride, RRL-09 (Selection)	420	200	0.50	Thymol (65-70%)
O. *basilicum*, RRL-011 (Improved strain)	500	220	0.45	Linalool (40%), Methyl chavicol (35%)
Synthesized Amphidiploidof *Ocimum gratissimum x O. viride*, RRL-015(New strain)	600	300	0.50	Eugenol (55-65%)
O. basilicum, CIM-Saumya	-	90	-	Methylchavicol (62%), Linalool (25%)
O. basilicum, CIM-Snigdha	-	-	-	Methylcinnamate (78.7%)
O. basilicum, CIM-Surabhi	-	110	-	Linalool (70-75%)
O. sanctum, CIM-Ayu	200	110	0.55	Eugenol (83%)
O. sanctum, CIM-Angna (Greyish purple stem with green leaves)	180	90	0.50	Eugenol (40%), Germacrene D (16%)

References

Balyan, S.S. and P. Pushpangadan. 1988. Pafai, 10(2): 13-19.

Getsadze, G.N. 1975. Referativnyl Zhurnal, 3: 55-64.

Khosla, M.K. 1995. Indian J. Genet., 55(1): 71-83.

Khosla, M.K. and S.N. Sobti, 1985. Cytologia, 50(2): 253-263.

Mario, R. Morales and James E. Simon. 1997. Hort. Science, 32(1): 148-149.

Mehra, P.N. and L.S. Gill. 1972. Cytologia, 27(1): 53-57.

Pushpangadan P., S.N. Sobti and Reayat Khan. 1975. The Nucleus, 18: 172-82.

Pushpangadan, P., S.N. Sobti, and R. Khan. 1975. Indian J. Exper. Biol., 13(5): 520.

Singh, T.P. 1978. Curr. Sci., 47(23): 915-916.

Snegirev, D.P. 1936. Bull. Applied Bot. Genet. and Plant Breeding. (USSR). Ser.111, 15: 245-273.

Verma, A. 1947. Arch. Soc. of Zool. Bot. Fennical "Vanamo", 2: 55-59.

Verma, P.K., M.S. Punia, G.D. Sharma and G. Talwar. 1989. Indian Perfumer, 33(2): 79-83.

Verma, P.K., S.N. Gupta, M. Khabiruddin and G.D. Sharma. 1998. Indian Perfumer, 42(1): 36-38.

Verma, P.K., Ravinder Singh, S.N. Gupta, O.P. Yadav and C.S. Tyagi. 2005. In: Course compendium of Winter School on "Advances in Medicinal, Aromatic and Under-utilized Plants Research" held at CCSHAU, Hisar from Sept. 29 - Oct.19, 2005.

Vij, S.P. and S.K. Kashyap. 1976. Cytologia, 41(3-4): 713-719.

Suggested questions

1. What is the scientific name of French basil, Holy basil, Mint basil, Long Tulsi and Kapoor Tulsi?
2. To which family Tulsi/Basil belongs.

3. What are the species belonging to *Basilicum* and *Sanctum* group along with their basic and somatic chromosome numbers?
4. What are the major chemical constituents of *Basilicum* and *Sanctum* group?
5. Name 3-4 variety developed by CIMAP, Lucknow.
6. What are the best suited methods for improvement of Ocimum species?
7. What is the significance of RRL-015 new strain of Tulsi?

23

Vetiver/Khas (*Chrysopogon zizanioides*)

Introduction

Vetiver/Khas is a fibrous, perennial and aromatic root crop spread over Indian sub-continent. The genus *Vetiveria* is derived from its Tamil name 'Vettiver'. It occurred largely in wet and damp environments over marshy lands in south-east Rajasthan and Haryana. Also found in river banks of western, central and northern U.P. up-to 1200 m elevation in the Himalayan foot-hills, T.N., Karnataka, Kerala and some parts of central India. Besides India, it is also widely grown in Sri Lanka, Indonesia, Haiti, Vietnam and Brazil for its fragrant volatile oil (Guenther, 1972). This species along with its unique xerophytic characteristics of tolerating prolonged drought, also survives under long seasonal flooding; extreme temperature and sustainability over a wide range of soil salinity and alkalinity. The South Indian types are rich in oil percent, but content more of hydrocarbons and less oxygenated constituents, thus inferior to North Indian types having higher percentage of oxygenated compound and very high quality odour value. It belongs to family Graminae or Poaceae.

Scientific Name	**Oil colour**	**Oil odour**	**Majorquality parameter**
Chrysopogon zizanioides	Viscous,light brown-dark brown	Sweet,arthy, and woody	Vetiverol content (45-85 %)

Uses

- Spongy and fibrous roots are commonly used for making cooler pads, floor mats, door screens emitting sweet fragrance with cooling effect in hot summer season when sprinkled with water as roots retain its delightful odour for a long period.

- The upper aerial parts with stem and inflorescence are greatly used for thatching of roofs.
- A highly fragrant viscous volatile oil is distilled from its roots, used in perfumery, cosmetic, toiletries and soap industries.
- The oil is carminative, hence useful in flatulence, colic pain and making of sherbet as cooling drink for summer.
- The roots are considered stimulant, diaphoretic and refrigerant in the Indian System of Medicine (Ayurved).
- It is a good hedge plant, as its fibrous roots stabilize and bind eroded sloppy lands.
- Roots travel long distance in light sandy soil in search of moisture, it is good for planting over river banks for making bunds, and also useful in soil and water conservation by protecting the soil erosion.
- The plants are not eaten by animals and rodents, except in early stage and during period of drought. Therefore, can also be used for fencing the fields.

Genotypic variability

It comprised of 10 species distributed in Tropical Asia, Africa and Australia. But besides *Chrysopogon zizanioides*, it has another Indian species, namely, *Vetiveria lawsonii*, which has a limited area in eastern Himalayas and Mysore in western peninsular India. But it has no commercial value in essential oil industry because of its poor oil quality. The only commercially grown species is *Vetiveria zizanioides,* which is known for its high quality root oil and highly desirable by perfumery industry since long.

Cytogenetics

Vetiver is an obligated cross pollinated aromatic grass with 84 percent hermaphrodite florets. It has been observed that seeds are not produced when selfed/bagged, but the puff collected from open pollinated panicle contain viable seeds. Pollen fertility is very high in North Indian types as compared with South Indian types.

Lavania (1985) studied chromosome morphology and nuclear DNA amount in 20 different collections ov vetiver and reported that the diploid somatic chromosome numbers were 2n=20.

In general, the karyotypes are nearly symmetrical and chromosome having median to sub-median centromere.

Later on **tetraploids (2n=40)** were developed using colchicine as duplicating agent, resulting in auto-tetraploid variety **'Sugandha'** by CIMAP, Lucknow.

Variability, Heritability and Genetic Advance

A considerable high amount of natural variability among different morphological traits including oil content, oil yield and root yield were recorded by various workers.

References	Heritability (%)	Genetic Advance (%)	Traits
Lal *et al.,* 2000	>90	49.5	For morphological traits
Punia *et al.,* 1989	95.27	54.55	Oil content
	87.36	51.77	Oil yield
	73.37	19.66	Root yield

Comparison of oil content in North Indian and South Indian Vetiver

Vetiver type	Range (Oil content- %)	Reference
North Indian Type	0.6-0.9	Dhingra *et al.,* 1954
South Indian Type	0.8-1.1	
Overall	0.6-1.1	Sethi *et al.*, 1978

Correlation and Path analysis

Reference	Characteristics
Ramanujam and Kumar, 1963a	Plant height, tiller number and lamina length showed positive and high correlation with root yield and may, therefore, serve as a good indication of root production. However, root yield was negatively correlated with oil content.
Lal *et al.*, 2000	Oil yield had positive and significant genotypic correlations with plant height, fresh root yield and dry root yield.
Lal *et al.*, 2001	Oil yield had positive and significant genotypic correlations with plant height, root length and oil content.
Gurnani *et al.*, 2004	Oil yield had positive and significant genotypic correlations with tiller number/plant, root yield/plant and root length. High positive direct effects were exhibited by plant height, tiller number/plant, oil content and vetiverol content on root yield/plant. Oil content also exhibited positive and strong direct effects on oil yield.

Breeding Strategies

The commercial cultivation of khas/vetiver depends upon high root yield and higher percentage of essential oil content in roots. Keeping this in view, past work on variability and association analysis by various workers found that improvement in the root yield and essential oil content is of prime importance. This can be achieved by adopting following breeding strategies.

i. **Clonal Selection:** Clonal selection is one of the most promising and economic breeding method in vegetative propagated crop. Lal *et al.* (1998) conducted varietal selection among 45 available accessions through repeated evaluation coupled with clonal selection for high root yield, oil content and oil yield. They further found one of the elite clone, BDP-1 to maintain its consistent superiority for oil yield. Subsequently, Lal *et al.* (1999) were able to develop three varieties, namely, Dharini, Gulabi and Kesari by repeated clonal selection at CIMAP, Lucknow. These produced high essential oil yield, but differing in aroma of oil. Dharini possessed the khus note and high root volume (longer and thicker roots); good soil binding ability useful for soil and water conservative; where as Gulabi and Kesari had rose and saffron like notes, respectively.

ii. **Polyploidy:** Another variety 'Sugandha' was developed at CIMAP, Lucknow with tetraploid i.e. 2n=40 chromosomes. This was isolated from materials introduced from Bharatpur (Rajasthan). The seedlings were subjected to polyploidy breeding, inducing auto-tetraploidy, purified and stabilized.

    Lavania (1993) conducted a trial on artificial auto-tetraploidy of vetiver taken in conjugation with source diploids and two improved diploids. The results revealed the superiority of auto-tetraploid strain 'Sugandha' over its diploid parent, exhibiting 60 percent improvement in oil productivity.

iii. **Inter-specific hybridization**: Sethi (1982) carried out a breeding programme at NBPGR, Delhi and selected 11 genotypes representing North India; and two of cultivated stock ex. South India, out of a total collection of 78 germplasm lines from all over the country.

The purity of these accessions were maintained through clonal multiplication. From segregating materials, 30 hybrid clonal progenies were fixed up. These were evaluated in preliminary trial, resulting in 12 promising hybrid clones, yielding better than local check for both root yield and essential oil content.

Study on stability parameters showed that Hybrid-26, Hybrid-6, Hybrid-7, Hybrid-8 and Hybrid-16 gave an overall higher root yield with 14.58, 13.54, 13.10, 12.10 and 9.44 q/ha; and oil content of 1.50, 1.20, 1.56, 1.85 and 1.55 per-cent, respectively. These values were better than general mean (10.14 q/ha and 1.44% oil content), and were significantly superior to local control (7.55 q/ha and 1.10%).

The Hybrid-8 (1.87%) and Hybrid-7 (1.60%) gave high oil content over normal soil at several locations at Delhi and Indore; and retained their superiority under marginal soils (saline alkaline), where oil content was estimated as 1.73% and 1.51%, respectively (Gupta *et al.,* 1983).

Sethi *et al.,* (1987) reported that Hybrid-8 was superior in odour value and then it was released as a variety

Pareek *et al.* (1991c) reported that there was progressive increase in oil content up-to 18 months under good management conditions and declined further with age.

Varietal achievements

Variety	Institute	Characters
Hybrid-8	NBPGR, New Delhi (1987)	**Diploid** hybrid, Tall growing (2m), perennial herb, high oil content (1.05% - DWB), Total vetiverol content (70-85%) and free vetiverol (60-70%)
Sugandha	CIMAP, Lucknow (1993)	**Auto-tetraploid** with 62% higher oil yield over diploid parent from Bharatpur.
Nilambore	Popular South Indian	High root yield (up to 38 q/ha, FWB), Oil yield (13-14 kg/ha)
Dharni	CIMAP, Lucknow	Clonal Selection with khus notes
Gulabi	CIMAP, Lucknow	Clonal Selection with rose like notes
Kesari	CIMAP, Lucknow	Clonal Selection with saffron like notes
KS-1	CIMAP, Lucknow	Oil yield: 20-25 kg/ha (1 year old)
CIM-Khusinalika	CIMAP, Lucknow	6 -8 months old crop
CIM - Vridhi	CIMAP, Lucknow	Suitable for drought and marginal lands

Active chemical constituents of oil

Major Chemical constituents	Quantity (%)
Khusimol	13.4-27.9
Vetiselinenol	10.3-19.5
Vetiverol	6.1-7
Beta-eudesmol	5.5-6
q-vetiver	1.5-2.5
Beta-vetiver	1.5-1.8

Besides, minor chemical constituents are Khusilol, Khusene, Khushol, Elemol and Vetivere.

Indian Standard parameters for oil of Vetiver are as under

Parameters	Value
Colour	Light to reddish brown, sometimes greenish, viscous liquid
Odour	Characteristic and persistent aroma with pleasant earthy woody balsamic character
Sp. Gravity (30°C)	0.990 to1.032
Optical rotation	50 to 130
Refractive index (30°C)	1.512 to 1.523
Acid value	Max. 40

Ester value	28 to 80
Ester value after acetylation	145 to 200
Carbonyl value	Min. 24
Total alcohols	Min. 70
Solubility	Soluble in 2 volumes of 80 per-cent ethyl alcohol

References

Dhingra, D.R., G.N. Gupta and Ganesh. 1954. In: Proceedings of Symposium on "R&D in Essential Oil Industries in India". pp. 17-22.

Guenther, E. 1972. The essential oils, Vol. 4; Robert E. Krieger Publishing Co. Inc. Huntington, New York.

Gupta, R.S., K.C. Trivedi, S. Verma and S.K. Gangrade. 1983. Indian Perfumer, 27(2):104-107.

Gurnani, D., D.K. Sharivastava, N. I. Dhani and A. Holkar. 2003. J. of Med. and Arom. Plants Sciences, 25(4): 959-964.

Lal, R.K. 2000. J. of Spices and Aromatic Crops, 9(2): 133-136.

Lal, R.K. and J.R. Sharma. 2002. J. of Med. and Arom. Plants Sciences, 22(18): 572-578.

Lal, R.K., J.R. Sharma, A.A. Naqui, and H.O. Mishra.1998. J. of Med. and Arom. Plants Sciences, 20(4): 1067-1070.

Lavania, V.C. 1985. Cytologia, 50: 177-185.

Lavania, V.C. 1991. J. of Essential Oil Research, 3(6): 455-457.

Pareek, S.K., M.L. Maheshwari and R. Gupta. 1991c. In: Symposium on "New Trends in Essential Oils". Tata Mc Grow-Hill Publishing Co. Ltd., pp. 269-279.

Punia, M.S., P.K. Verma and G.D. Sharma. 1989. Indian J. of Agricultural Science, 59(1): 71-72.

Ramanujam, S., Sushil Kumar. 1963a. Indian J. Genet., 23(1): 82-89.

Sethi, K.L. 1982. Indian Perfumer, 26(2-4): 54-61.

Sethi, K.L., M.L. Maheshwari and R. Gupta. 1987. Hybrid-8, New khas variety. Indian Fmg., 36(12): 13-17.

Verma, P.K., Ravinder Singh, S.N. Gupta, O.P. Yadav and C.S. Tyagi. 2005. In: Course Compendium of Winter School on "Advances in Medicinal, Aromatic and Under-utilized Plants Research" held at CCS HAU, Hisar (Haryana) from Sept. 29 - Oct.19, pp. 54-62.

Suggested questions

1. What is the popular scientific name of Vetiver/Khas?
2. To which family Vetiver/Khas belongs.
3. What is the name of other Vetiver species?
4. What are the major chemical constituents of *Vetiver* oil?
5. Name 3-4 variety developed by CIMAP, Lucknow.
6. Name the only polyploidy variety developed by CIMAP, Lucknow along with its type of ploidy level and chromosome number.
7. Name the varieties developed by clonal selection and hybridization.

24

Wild Marigold (*Tagetes minuta* L.)

Introduction

Wild marigold (*Tagetes minuta*) is also famous as African marigold and mint marigold, valued for its higher content and superior in essential oil quality among all the cultivated species undergenus*Tagetes*. Marigold/Genda (*Tagetes* species) are mostly grown for scented flowers for utilization in functions all over the world. The flowers and leaves are of different varieties, which are identified by their distinct odouriferous oil.

Origin

It is inhabitant to the grasslands of temperate region in South America and now become widespread throughout the world. In India, it has been customized in the North-West Himalayas up-to an altitudes ranging from1000-2500 meters in H.P., U.P. and J&K. Now-a-days, *Tagetes minuta* is considered as an extensive weed in South Europe, Africa, South Asia, Australia, and Madagascar (Chalchat *et al*., 1995). It is found growing naturally as a noxious weed in the wild and arable farming systems.

The high demand of its essential oil may leads into reduction of the wild resources, thus it is essential to start systematic cultivation in sub-tropical to temperate agro-climatic zones all over the world, mainly in the region of Indian subcontinent (Singh *et al*., 2003).

Distribution

Tagetes minuta can be grown successfully at both low and high altitudes in low or high rainfall areas. It is also noticed in disturbed areas during early development stages, as seeds are disseminated through wild and domesticated animal hairs and human clothes. But by the finish of its growing season, aerial parts dried and easily destroyed by fire. The seeds deposition in the soil leads to development of new colonies in the following season. It is pioneer plant in South Africa and specially recommended of disturbed land for its reclamation. The plant is wide spread in adjacent to forest margin and forest roadside; and dominating arable land as the common weed after crop harvest. It can sustain soil pH from 4.3 to 7.0.

Uses

- In South America, mint marigold is utilised as a spice giving a flavour resembling to apple.
- Its fresh as well as dried leaves are useful as an aromatic flavour for vegetables and soups.
- In some parts of Peru, it is grown as vegetable, where dried leaves are used as flavouring agent and condiments in different types of food items.
- Essential oil is also extracted from leaves and floral heads for utilization as a flavouring agent in most of the food items, including candy, frozen dairy desserts, baked goods, cola beverages, alcoholic beverages, puddings, gelatines, relishes and condiments.
- Chopped fresh leaves used to season green salads, chicken, anise-flavoured tea and brew a sweet.
- Mint marigold is also used to flavour liquors.
- The strong-smelling essential oils of *T. minuta* enabled it for many purposes, including diuretic, laxative, insect repellent, flavouring, snuf and stimulant.
- It is also utilizsed for the treatment of stomach cramps coughs, and rheumatism.
- It is cultivated for the perfumery industry in South Africa (Gil *et al.*, 1996).
- It has been reported to act as nematicide and herbicide both.

Nematicidal and herbicidal activity

T. minuta is found strong in nematicidal and insecticidal properties, both as a fresh plant and when distilled provide evidence for use against nematodes. Effects of *T. minuta* were found better than the *T. erecta* and *T. patula*. However, achievement has been uneven and result depends on environmental conditions and several other factors.

Roots secretion has herbicidal properties, and inhibits the growth of some plants growing nearby. It is also effective to control some perennial weeds, such as *Aegopodium podagraria* (ground elder), *Glechoma hederacea* (ground ivy), *Ranunculus ficaria* (celandine), *Convolvulus arvensis* (field bindweed and *Agropyron repens* (couch grass) (Ravindran, 2017). Generally, fresh herbs are placed in the rooms to keep away cockroaches (Beccaloni, 1991).

Genotypic variability

Soule(1996) reported that genus *Tagetes* comprised of twenty seven (27) annual and twenty nine (29) perennial species; mostly from sub-tropical America and one from tropical Africa. The genus name derived from Latin name for marigold, 'Tagetes', and *minuta* means very small individual flowers within the inflorescences.

The genus *Tagetes* comes under Compositae family, and has 40 species divided into two subgenera *viz.*, subgenus *Cucida* and *Tagetes*. Four species under genus *Tagetes* are reported to found in India:

- *T. minuta* L.
- *T. patula,*
- *T. erecta*
- *T. signata.*

Cytogenetics

T. minuta L. is a highly aromatic annual herb/shrub growing wild and cultivated for ornamental flowers. It has been domesticated in India, but no specific variety has been developed so far. Variations for morphological characters has been observed in plant populations, indicating scope for selection and crop improvement. In general, higher ratio of flowers and leaves to that of stem is required for higher oil yield.

No study with regard to chromosome number has been reported for *Tagetes minuta*.

Floral biology

- *Tagetes minuta* (mint marigold) plants are erect, woody annual herbs take growth up-to 2 m height with sturdily odorous foliage.
- Its taproot is generally short tapering encircled by several fibrous roots, and also has mycorrhizal links.
- Its stem is ridged or grooved, initially it is green but at maturity turns to brownish / reddish; and bears branches in the upper portion unless cut off / broken near the base.
- It generally bears leaves in opposite pairs on the main stem and but singly on lateral branches.

- Its leaves are slightly glossy-green, pinnately compound, varied from 5to 20 cm long, and having 4–6 pairs of leaflets. Its leaflets are sharply toothed with 2–4 cm long and thin lanceolate.
- The dorsal part of the leaf has a number of orange, small, punctate, multicellular glands, which emanate a liquorice-like aroma when burst out.
- Glands are found situated along the midrib and towards the margins of leaflets, and also to be noticed on the stem and involucre bracts.
- The panicle-like scented inflorescences, composed of 20–80 scarcely cylindrical flower heads.
- The flower heads are small, about 1.0–1.5 cm long and 0.3–0.4 cm in diameter, bounded by 4–5 fused involucre bracts, and spotted with glands remain associated till the maturity.
- Ray and disk florets barely extend beyond the phyllaries; ray florets are 3–5 in number and yellow-orange in colour, while disk florets are 10–15 in number but also yellow-orange in colour (Soule, 1996).
- Its seed is known achenes and having dark-brown colour, 10–20 mm long, and associated with a pappas of 1–4 tiny scales and 1–3 mm long.
- Its mature seeds are 5–8 mm long, cylindrical, black with fine hairs, and tipped with a chaffy pappus of 4–8 scales.
- Its seedlings are epigeal in germination.

The slender, often reddish, hypocotyls are 10–15 mm long, and terminate in elongated cotyledons approximately 1 cm long, which are often reddish in lower side. The first pair of leaves is intensely divided into three parts with one big terminal lobe and has unevenly toothed edges.

Reproductive biology

- *T. minuta* plants generally bears relatively small seeds, without any kind of dormancy and has 7–8 months' germination longevity.
- It is propagated by seeds, which germinate over a period of 48 h; most seeds germinate at 20 - 30°C (optimal 25°C).
- Fresh seeds germination was observed up to 95 percent.
- There is no requirement of sunlight for germination of seeds; however, they counter it very positively, so that germination only occurs from seeds near the soil surface upto 6 mm.

Breeding Methodoloy

No work has been reported so far in literature with regard to breeding methodology for crop improvement work.

Varietal achievement

T. minuta var. *vanphool*, derivative of an open population variety cultivated in northern part of India and has a high oil yield and quality (Kumar *et al.*, 1999).

Chemical composition of oil

The chemical composition of essential oil obtained from *T. minuta* received from market consists of three major ketones (80-90 % oil) *viz.*, dihydrotagetone, tagetone and ocimenone.

The young crop in vegetative state has major amount of tagetone, but with the maturity of the crop, the dihydrotagetone increases with consequent decrease in tagetone and ocimenone contents.

At full bloom stage, when yield of both herb and its oil are maximum, the three ketones present in the oil are in the ratio of dihydrotagetone (45-60%), tagetone (15-25%), and ocimenone (20-30%). The said proportion of ketones contributes its good quality oil for use in the industry.

Oil storage

The storage of oil induces major changes in the compositions of three ketones in the oil. In general, the presence of ocimenone is not found in freshly distilled oil, but it makes its presence after a period of 15 days of storage, possibly due to the polymerisation. Therefore, it is advisable that the storage of its oil should be done in amber-coloured bottles, aluminium/SS containers free from air and moisture to maintain its quality.

Reference

Singh Virendra, R.P. Singh, Bikram Singh and V.K. Kaul. 1995. In: Advances in Horticulture. Vol. 11, Ed. by K.L. Chadha and Rajendra Gupta. Malhotra Publishing House, New Delhi, pp. 789-793.

Suggested questions

1. What is the scientific name of Wild marigold?
2. To which family Wild marigold belongs.
3. What are the other three species of Wild marigold?
4. What are the chemical constituents of Wild marigold oil?
5. Name the only variety developed.
6. What is the proper stage of harvest when all the three chemicals are present in the oil?

Index

Aromatherapy 1, 16, 32, 37, 38, 39, 41, 42, 90, 98, 127Disease Resistance 47, 48, 93